3rd Workshop on Theory, Modelling and Computational Methods for Semiconductors 2012

(TMCSIII)

AA008917

Journal of Physics: Conference Series Volume 367

Leeds, United Kingdom
18-20 January 2012

Editors:

Marco Califano
Max Migliorato
Matt Probert

ISBN: 978-1-62276-784-7
ISSN: 1742-6588

Printed from e-media with permission by:

Curran Associates, Inc.
57 Morehouse Lane
Red Hook, NY 12571

CURRAN ASSOCIATES INC. proceedings.com

Some format issues inherent in the e-media version may also appear in this print version.

Printed by Curran Associates, Inc. (2013)

For permission requests, please contact the Institute of Physics at the address below.

Institute of Physics
Dirac House, Temple Back
Bristol BS1 6BE UK

Phone: 44 1 17 929 7481
Fax: 44 1 17 920 0979

techtracking@iop.org

Additional copies of this publication are available from:

Curran Associates, Inc.
57 Morehouse Lane
Red Hook, NY 12571 USA
Phone: 845-758-0400
Fax: 845-758-2634
Email: curran@proceedings.com
Web: www.proceedings.com

TABLE OF CONTENTS

AB INITIO

NITRIDES

QUANTUM NANOSTRUCTURES

DEVICES

NANOSCALE SCIENCE AND DESIGN

Journal of Physics: Conference Series **367** (2012) 011001 doi:10.1088/1742-6596/367/1/011001

Preface

These conference proceedings contain the written papers of the contributions presented at the 3rd International Conference on Theory, Modelling and Computational Methods for Semiconductor materials and nanostructures. The conference was held at the School of Electronic and Electrical Engineering, University of Leeds, Leeds, UK on 18–20 January 2012. The previous conferences in this series took place in 2010 at St William's College, York and in 2008 at the University of Manchester, UK.

The development of high-speed computer architectures is finally allowing the routine use of accurate methods for calculating the structural, thermodynamic, vibrational, optical and electronic properties of semiconductors and their hetero- and nano-structures. The scope of this conference embraces modelling, theory and the use of sophisticated computational tools in semiconductor science and technology, where there is substantial potential for time-saving in R&D. Theoretical approaches represented in this meeting included: Density Functional Theory, Tight Binding, Semiempirical Pseudopotential Methods, Effective Mass Models, Empirical Potential Methods and Multiscale Approaches. Topics included, but were not limited to: Optical and Transport Properties of Quantum Nanostructures including Colloids and Nanotubes, Plasmonics, Magnetic Semiconductors, Graphene, Lasers, Photonic Structures, Photovoltaic and Electronic Devices. This workshop ran for three days, with the objective of bringing together UK and international leading experts in the theoretical modelling of Group IV, III-V and II-VI semiconductors, as well as students, postdocs and early-career researchers. The first day focused on providing an introduction and overview of this vast field, aimed particularly at students, with several lectures given by recognised experts in various theoretical approaches. The following two days showcased some of the best theoretical research carried out in the UK in this field, with several contributions also from representatives of renowned theoretical groups from many European countries (Spain, France, Ireland, Germany, Italy, Poland, Denmark, Sweden, Serbia, Greece, etc.), as well as Asia (India) and Africa (Algeria, Tunisia and South Africa).

We would like to thank all participants for making this a very successful meeting and for their contribution to the conference programme and these proceedings. We would also like to acknowledge the financial support from the Institute of Physics (Computational Physics group and Semiconductor Physics group), and QuantumWise (distributors of Atomistix).

The Editors

Acknowledgments

Conference Organising Committee:
Marco Califano (University of Leeds)
Max Migliorato (University of Manchester)
Matt Probert (University of York)

Journal of Physics: Conference Series **367** (2012) 011001 doi:10.1088/1742-6596/367/1/011001

Programme Committee:
Stewart Clark (University of Durham)
Aldo Di Carlo (University of Rome 'Tor Vergata', Italy)
Ben Hourahine (University of Strathclyde)
Lev Kantorovich (King's College London)
Risto Nieminen (Helsinki University of Technology, Finland)
Eoin O'Reilly (Tyndall Institute Cork, Republic of Ireland)
Mauro Pereira (Sheffield Hallam University)
John Robertson (University of Cambridge)
Mervin Roy (University of Leicester)
Stanko Tomic (University of Salford)
David Whittaker (University of Sheffield)

The proceedings were edited and compiled by Marco Califano, Max Migliorato and Matt Probert.

Journal of Physics: Conference Series **367** (2012) 011002 doi:10.1088/1742-6596/367/1/011002

Peer review statement

All papers published in this volume of *Journal of Physics: Conference Series* have been peer reviewed through processes administered by the proceedings Editors. Reviews were conducted by expert referees to the professional and scientific standards expected of a proceedings journal published by IOP Publishing.

Journal of Physics: Conference Series **367** (2012) 012001 doi:10.1088/1742-6596/367/1/012001

Optical response of extended systems from time-dependent Hartree-Fock and time-dependent density-functional theory

Leonardo Bernasconi[1], Ross Webster[2], Stanko Tomić[3] and Nicholas M Harrison[1,2]

[1] STFC Rutherford Appleton Laboratory, Harwell Oxford, Didcot, OX11 0QX, UK
[2] Department of Chemistry, Imperial College London, London SW7 2AZ, UK
[3] Joule Physics Laboratory, University of Salford, Manchester, M5 4WT, UK

E-mail: leonardo.bernasconi@stfc.ac.uk

Abstract. We describe a unified formulation of time-dependent Hartree-Fock (TD-HF) and time-dependent density-functional theory (TD-DFT) for the accurate and efficient calculation of the optical response of infinite (periodic) systems. The method is formulated within the linear-response approximation, but it can easily be extended to include higher-order response contributions, and, in TD-DFT, it can treat with comparable computational efficiency purely local, semi-local or fully non-local approximations for the ground-state exchange-correlation (XC) functional and for the response TD-DFT XC kernel in the adiabatic approximation. At variance with existing methods for computing excitation energies based on the diagonalisation of suitable coupling matrices, or on the inversion of a dielectric matrix, our approach exploits an iterative procedure similar to a standard self-consistent field calculation. This results in a particularly efficient treatment of the coupling of excitations at different **k** points in the Brillouin zone. As a consequence, our method has the potential to describe completely from first principles the optically induced formation of bound particle-hole pairs in wide classes of materials. This point is illustrated by computing the optical gaps of a series of representative bulk semiconductors, (non-spin polarised) oxides and ionic insulators.

1. Introduction

The development of accurate methods to account, fully from first principles, for the response of a quantum system of nuclei and electrons to an electromagnetic radiation of optical or UV frequency is currently a very active and complex area of research in solid state physics and materials science. Technologically, predictive models of the optical response of extended many-electron systems offer a potentially very powerful route to the design and optimisation of new materials of relevance *e.g.* in photo-voltaics and solar light conversion, photo-catalysis, and quantum control of single atoms. Theoretically, the *ab initio* description of electronic excitations provides means to study fundamental issues in the quantum theory of condensed phases, including, for instance, the effects of electron correlation in excited states *and* in the ground state of a many-electron system.

TD-DFT [1] has emerged in recent years as a powerful and general approach to address the computation of several excited state properties, including excitation energies, transition probabilities, and excited state energy gradients with respect to nuclear displacements. [2–6]

Journal of Physics: Conference Series **367** (2012) 012001 doi:10.1088/1742-6596/367/1/012001

TD-DFT is formally an exact theory of the time-evolution of a many-electron system in the presence of an external time-dependent perturbation. [7] As such, it allows one to study the real-time evolution of an interacting ensemble of electrons perturbed by the presence of the external field.

In the limit of a weak field, *e.g.* in the simulation of absorption spectra, TD-DFT can be formulated using a particularly elegant formalism [8, 9] that yields the interacting linear density response function $n^{(1)}(\mathbf{r}'', \omega)$ in terms of the non-interacting (Kohn-Sham [10, 11]) response function $\chi_s(\mathbf{r}, \mathbf{r}'; \omega)$ *via* a Dyson-like equation,

$$\begin{aligned}&\int d\mathbf{r}' \chi_s(\mathbf{r}, \mathbf{r}'; \omega) v^{(1)}(\mathbf{r}', \omega) = \\ &\int d\mathbf{r}'' \left[\delta(\mathbf{r} - \mathbf{r}'') - \int d\mathbf{r}' \chi_s(\mathbf{r}, \mathbf{r}'; \omega) \left\{ \frac{1}{|\mathbf{r}' - \mathbf{r}''|} + f^{\mathrm{xc}}(\mathbf{r}', \mathbf{r}''; \omega) \right\} \right] n^{(1)}(\mathbf{r}'', \omega).\end{aligned} \tag{1}$$

$v^{(1)}(\mathbf{r}', \omega)$ describes an external field oscillating at a frequency ω, and $f^{\mathrm{xc}}(\mathbf{r}, \mathbf{r}'; \omega)$ is the frequency representation of the XC kernel $f^{\mathrm{xc}}(\mathbf{r}, t; \mathbf{r}', t')$, describing the response of the XC component of the Kohn-Sham potential to the external perturbation,

$$f^{\mathrm{xc}}(\mathbf{r}, t; \mathbf{r}', t') = \left. \frac{\delta^2 A^{\mathrm{xc}}[n](\mathbf{r}, t)}{\delta n(\mathbf{r}, t) \delta n(\mathbf{r}', t')} \right|_{n(\mathbf{r},t)=n^{(0)}(\mathbf{r})} = \left. \frac{\delta v^{\mathrm{xc}}[n](\mathbf{r}, t)}{\delta n(\mathbf{r}', t')} \right|_{n(\mathbf{r},t)=n^{(0)}(\mathbf{r})}, \tag{2}$$

with $A^{\mathrm{xc}}[n](\mathbf{r}, t)$ being the XC action functional, and $\delta/\delta n(\mathbf{r}, t)$ indicating a functional derivative with respect to $n(\mathbf{r}, t)$. The exact form of $f^{\mathrm{xc}}(\mathbf{r}, t; \mathbf{r}', t')$ is, in general, unknown, and this quantity, similarly to the ground state XC functional of density-functional theory (DFT), has to be approximated. Frequently, practical applications of TD-DFT employ the adiabatic approximation, [12] in which the frequency dependence of the kernel is neglected and the functional derivative of the XC potential in Eqn. (2) replaced by the second functional derivative of an XC energy, *e.g.* the Local Density Approximation (LDA) XC energy, with respect to the density response. This leads to a purely local and frequency independent kernel. The poles of $n^{(1)}(\mathbf{r}'', \omega)$ correspond to the *many-body* excitation energies of the system, whose modes couple with the external field.

TD-DFT excitation energies can also be computed using an alternative procedure (which we will indicate here as the *RPA-matrix formalism*) based on the solution of a generalised anti-Hermitian eigenvalue equation, [13, 14]

$$\begin{pmatrix} \mathbf{A} & \mathbf{B} \\ \mathbf{B}^* & \mathbf{A}^* \end{pmatrix} \begin{pmatrix} \mathbf{X} \\ \mathbf{Y} \end{pmatrix} = \omega \begin{pmatrix} -\mathbf{1} & \mathbf{0} \\ \mathbf{0} & \mathbf{1} \end{pmatrix} \begin{pmatrix} \mathbf{X} \\ \mathbf{Y} \end{pmatrix}. \tag{3}$$

The eigenvalues of this equation correspond to the free oscillations of the system, and the coupling with the external field is described by the left and right eigenvectors $\mathbf{X}$ and $\mathbf{Y}$, which yield the corresponding (possibly vanishing) transition intensities. The matrices $\mathbf{A}$ and $\mathbf{B}$ describe the coupling of independent-particle Kohn-Sham excitations $a \leftarrow i$ and $b \leftarrow j$ (with $i, j, \ldots$ and $a, b, \ldots$ labelling occupied and unoccupied Kohn-Sham orbitals respectively) *via* Coulomb and XC interaction,

$$\begin{aligned} A_{ai,bj} &= \delta_{ab}\delta_{ij}(\varepsilon_a - \varepsilon_i) + (ai|jb) + (ai|f_{\mathrm{xc}}|jb) \\ B_{ai,bj} &= (ai|bj) + (ai|f_{\mathrm{xc}}|bj). \end{aligned} \tag{4}$$

The two-electron integrals $(pq|rs)$ are in Mulliken's notation, [15] and the XC kernel integrals are given by

$$(ai|f_{\mathrm{xc}}|bj) = \int d\mathbf{r} d\mathbf{r}' \phi_a^*(\mathbf{r}) \phi_i(\mathbf{r}) f^{\mathrm{xc}}(\mathbf{r}, \mathbf{r}'; \omega) \phi_b^*(\mathbf{r}') \phi_j(\mathbf{r}'). \tag{5}$$

Journal of Physics: Conference Series **367** (2012) 012001 doi:10.1088/1742-6596/367/1/012001

This approach is formally analogous to the TD-HF method, [16] which leads to an eigenvalue equation identical to Eqn. (3) with the matrices **A** and **B** now given by

$$\begin{aligned} A_{ai,bj} &= \delta_{ab}\delta_{ij}(\varepsilon_a - \varepsilon_i) + (ai|jb) - (ab|ji) \\ B_{ai,bj} &= (ai|bj) - (aj|bi). \end{aligned} \tag{6}$$

Eqn. (4) and (6) can be therefore be combined to give the *hybrid* TD-DFT coupling matrices

$$\begin{aligned} A_{ai,bj} &= \delta_{ab}\delta_{ij}(\varepsilon_a - \varepsilon_i) + (ai|jb) - c_{\mathrm{HF}}(ab|ji) + (1 - c_{\mathrm{HF}})(ai|f_{\mathrm{xc}}|jb) \\ B_{ai,bj} &= (ai|bj) - c_{\mathrm{HF}}(aj|bi) + (1 - c_{\mathrm{HF}})(ai|f_{\mathrm{xc}}|bj), \end{aligned} \tag{7}$$

where c_{HF} measures the fraction of Hartree-Fock exchange in the hybrid response (*e.g.* $c_{\mathrm{HF}} = 0.2$ for TD-DFT at the B3LYP [17] level of theory). For infinite periodic systems, the Kohn-Sham orbitals have the Bloch form, $\langle \mathbf{r}|p\mathbf{k}\rangle = e^{i\mathbf{k}\cdot\mathbf{r}}u_{\mathbf{k}p}(\mathbf{r})$, with one-particle energies $\varepsilon_p^{\mathbf{k}}$, and Eqns. (7) have to be generalised, to give [18]

$$\begin{aligned} A_{ai,bj}^{\mathbf{k}_a\mathbf{k}_i,\mathbf{k}_b\mathbf{k}_j} &= \delta_{ij}\delta_{ab}\delta_{\mathbf{k}_i\mathbf{k}_j}\delta_{\mathbf{k}_a\mathbf{k}_b}(\varepsilon_a^{\mathbf{k}_a} - \varepsilon_i^{\mathbf{k}_i}) + (a\mathbf{k}_a i\mathbf{k}_i|j\mathbf{k}_j b\mathbf{k}_b) - c_{\mathrm{HF}}(a\mathbf{k}_a b\mathbf{k}_b|j\mathbf{k}_j i\mathbf{k}_i) \\ &\quad + (1 - c_{\mathrm{HF}})(a\mathbf{k}_a i\mathbf{k}_i|f_{\mathrm{xc}}|j\mathbf{k}_j b\mathbf{k}_b) \\ B_{ai,bj}^{\mathbf{k}_a\mathbf{k}_i,\mathbf{k}_b\mathbf{k}_j} &= (a\mathbf{k}_a i\mathbf{k}_i|b\mathbf{k}_b j\mathbf{k}_j) - c_{\mathrm{HF}}(a\mathbf{k}_a j\mathbf{k}_j|b\mathbf{k}_b i\mathbf{k}_i) + (1 - c_{\mathrm{HF}})(a\mathbf{k}_a i\mathbf{k}_i|f_{\mathrm{xc}}|b\mathbf{k}_b j\mathbf{k}_j). \end{aligned} \tag{8}$$

Eqn. (3) can then in principle be solved by iterative diagonalisation for a periodic system using the definitions given in Eqn. (8), to obtain an absorption profile from a discrete set of (several) many-body excitations. In practice, however, this approach rapidly becomes impractical as the size of the system increases, or if very fine meshes of **k** points are used to sample the Brillouin zone.

In this paper, we describe a formulation of TD-HF, TD-DFT and hybrid TD-HF/TD-DFT for weak perturbations that obtains excitation energies without resorting to a diagonalisation of a large coupling matrix, as in Eqn. (3), and that does not require the calculation (and subsequent inversion, to obtain excitation energies) of $\chi_s(\mathbf{r}, \mathbf{r}'; \omega)$, as in Eqn. (1). The method is closely related to the (frequency-dependent) coupled-perturbed Hartree-Fock/DFT theories [19–21] widely used in quantum chemistry and molecular physics, which we generalise here to infinite periodic systems. Our approach is formulated in the Linear Combination of Atomic Orbital (LCAO) approximation, with an all-electron basis set of Gaussian functions, as implemented in the `CRYSTAL` [22–24] code. This choice allows us to compute analytically and extremely efficiently the two-electron integrals $(pq|rs)$, and to arrive at an algorithm that scales linearly with the number of **k** points used to sample the Brillouin zone. [25]

The paper is organised as follows. In Section 2 we derive the working equations of our method, and we show how excitation energies can be determined from the poles of suitable response functions. We also show how the frequency-dependent dielectric tensor, and a number of related quantitites can be computed using our approach. In Section 3, we describe examples of the application of our method to the calculation of the optical gaps of a number of semiconductors and crystalline ionic systems, and show the level of accuracy achieved in each class of materials. We demonstrate that our method describes quantitatively weakly bound excitons in semiconductors, and, possibly, tighly bound excitons in crystalline materials. The more complex case of charge-transfer excitons in highly ionic systems is also briefly addressed.

2. Theory

2.1. Derivation of frequency-dependent coupled-perturbed Hartree-Fock and DFT equations

The equation of motion for the **k**-dependent Fock (or Kohn-Sham) matrix $\mathbf{F}^{\mathbf{k}}$ in the presence of an external time-dependent perturbation,

$$\mathbf{F}^{\mathbf{k}}\mathbf{C}^{\mathbf{k}} - i\frac{\partial}{\partial t}\mathbf{S}^{\mathbf{k}}\mathbf{C}^{\mathbf{k}} = \mathbf{S}^{\mathbf{k}}\mathbf{C}^{\mathbf{k}}\mathbf{E}^{\mathbf{k}} \tag{9}$$

can be derived from Frenkel's variational principle, [20, 26] subject to the orthonormality constraint

$$\frac{\partial}{\partial t}\mathbf{C}^{\mathbf{k}\dagger}\mathbf{S}^{\mathbf{k}}\mathbf{C}^{\mathbf{k}} = \mathbf{0}. \tag{10}$$

Throughout this derivation, atomic units for all quantities are assumed. Here, $\mathbf{C}^{\mathbf{k}}$ is a matrix of crystal orbital (CO) coefficients. The COs, $\psi_i(\mathbf{r},\mathbf{k})$, are combinations of N Bloch functions (BFs) $\phi_\mu(\mathbf{r},\mathbf{k})$,

$$\psi_i(\mathbf{r},\mathbf{k}) = \sum_{\mu}^{N} C^{\mathbf{k}}_{\mu i}\phi_\mu(\mathbf{r},\mathbf{k}). \tag{11}$$

The elements of the overlap matrix $\mathbf{S}^{\mathbf{k}}$ are given by

$$S^{\mathbf{k}}_{\mu\nu} = \int d\mathbf{r}\ \phi_\mu(\mathbf{r},\mathbf{k})\phi_\nu(\mathbf{r},\mathbf{k}), \tag{12}$$

and $\mathbf{E}^{\mathbf{k}}$ is the orbital energy matrix. The BFs are expressed in terms of a set of atom-centred Gaussian basis set functions, $\chi_\mu(\mathbf{r}-\mathbf{R})$,

$$\phi_\mu(\mathbf{r},\mathbf{k}) = N^{-\frac{1}{2}}\sum_{\mathbf{R}}^{N} e^{i\mathbf{k}\cdot\mathbf{r}}\chi_\mu(\mathbf{r}-\mathbf{R}), \tag{13}$$

where the sum runs on the real-lattice vectors $\mathbf{R}$, and $N \to \infty$.

In general, a time dependent perturbation of the Hamiltonian operator can be expressed as a sum of oscillatory contributions with frequencies $\pm\omega$

$$\boldsymbol{\lambda}_\omega = \boldsymbol{\lambda}_{+\omega}e^{+i\omega t} + \boldsymbol{\lambda}_{-\omega}e^{-i\omega t}, \tag{14}$$

$\boldsymbol{\lambda}_{\pm\omega} = (\lambda^x_{\pm\omega}, \lambda^y_{\pm\omega}, \lambda^z_{\pm\omega})$ are the electric field Cartesian components oscillating at frequencies $\pm\omega$. The coupling between the electrons and the external field is described by the perturbation Hamiltonian

$$\hat{H}' = \boldsymbol{\lambda}_{\pm\omega}\cdot\hat{\mathbf{r}}, \tag{15}$$

where $\hat{\mathbf{r}}$ is the position operator. For finite systems, this corresponds to the standard position operator $\hat{\mathbf{r}}^F = (x, y, z)$. For an infinite periodic system, $\hat{\mathbf{r}}^F$ is unbound, and it has to be replaced by its generalised, translational-invariant form [20, 27–29]

$$\hat{\mathbf{r}} = ie^{i\mathbf{k}\cdot\mathbf{r}}\nabla_{\mathbf{k}}e^{-i\mathbf{k}\cdot\mathbf{r}}. \tag{16}$$

We remark that using the definition of Eqn. (16) for $\hat{\mathbf{r}}$ is formally equivalent to introducing an electric field in the Hamiltonian *via* a vector potential $A(t)$, to replace the kinetic energy operator $\hat{p}^2/2$ with $|\hat{p}+A(t)|^2/2$. [20] Both approaches maintain the translational invariance of the Hamiltonian, and they are related by a gauge transformation. [20, 29] Ultimately, they can be shown to lead to the same set of coupled-perturbed equations.

The matrices $\mathbf{F}^{\mathbf{k}}$, $\mathbf{C}^{\mathbf{k}}$, $\mathbf{E}^{\mathbf{k}}$ and $\mathbf{S}^{\mathbf{k}}$ in Eqn. (9) can be expanded in the perturbative parameters $\boldsymbol{\lambda}_{\pm\omega}$, and, for $\boldsymbol{\lambda}_{\pm\omega} \to \mathbf{0}$, the linear response approximation for a pair of fixed frequencies $\pm\omega$ leads to

$$\begin{aligned}
\mathbf{F}^{\mathbf{k}}_\omega &= \underline{\mathbf{F}}^{\mathbf{k}} + \lambda^a_{+\omega}\mathbf{F}^{\mathbf{k},a}_{+\omega}e^{+i\omega t} + \lambda^a_{-\omega}\mathbf{F}^{\mathbf{k},a}_{-\omega}e^{-i\omega t},\\
\mathbf{C}^{\mathbf{k}}_\omega &= \underline{\mathbf{C}}^{\mathbf{k}} + \lambda^a_{+\omega}\mathbf{C}^{\mathbf{k},a}_{+\omega}e^{+i\omega t} + \lambda^a_{-\omega}\mathbf{C}^{\mathbf{k},a}_{-\omega}e^{-i\omega t},\\
\mathbf{E}^{\mathbf{k}}_\omega &= \underline{\mathbf{E}}^{\mathbf{k}} + \lambda^a_{+\omega}\mathbf{E}^{\mathbf{k},a}_{+\omega}e^{+i\omega t} + \lambda^a_{-\omega}\mathbf{E}^{\mathbf{k},a}_{-\omega}e^{-i\omega t},\\
\mathbf{S}^{\mathbf{k}}_\omega &= \underline{\mathbf{S}}^{\mathbf{k}} + \lambda^a_{+\omega}\mathbf{S}^{\mathbf{k},a}_{+\omega}e^{+i\omega t} + \lambda^a_{-\omega}\mathbf{S}^{\mathbf{k},a}_{-\omega}e^{-i\omega t},
\end{aligned} \tag{17}$$

where a represents the Cartesian component of the applied field, and $\underline{\mathbf{F}}^{\mathbf{k}}$, $\underline{\mathbf{C}}^{\mathbf{k}}$, $\underline{\mathbf{E}}^{\mathbf{k}}$ and $\underline{\mathbf{S}}^{\mathbf{k}}$ are the matrices solving the unperturbed Hartree-Fock (or Kohn-Sham) equation

$$\underline{\mathbf{F}}^{\mathbf{k}}\underline{\mathbf{C}}^{\mathbf{k}} = \underline{\mathbf{S}}^{\mathbf{k}}\underline{\mathbf{C}}^{\mathbf{k}}\underline{\mathbf{E}}^{\mathbf{k}}. \tag{18}$$

In Eqn. (17) we have defined the response matrices

$$\begin{aligned}
\mathbf{F}^{\mathbf{k},a}_{\pm\omega} &= \partial \mathbf{F}^{\mathbf{k}}/\partial\lambda^{a}_{\pm\omega} \\
\mathbf{C}^{\mathbf{k},a}_{\pm\omega} &= \partial \mathbf{C}^{\mathbf{k}}/\partial\lambda^{a}_{\pm\omega} \\
\mathbf{E}^{\mathbf{k},a}_{\pm\omega} &= \partial \mathbf{E}^{\mathbf{k}}/\partial\lambda^{a}_{\pm\omega} \\
\mathbf{S}^{\mathbf{k},a}_{\pm\omega} &= \partial \mathbf{S}^{\mathbf{k}}/\partial\lambda^{a}_{\pm\omega}
\end{aligned} \tag{19}$$

representing the gradients of the Fock/Kohn-Sham matrix, CO coefficients, orbitals energies and overlap matrix respectively to the external perturbation. Assuming that the nuclei are at fixed positions (*i.e.* that the electronic and nuclear motions are decoupled) leads to $\partial\mathbf{S}^{\mathbf{k}}/\partial\lambda^{a}_{\pm\omega} = \mathbf{0}$, *i.e.* $\mathbf{S}^{\mathbf{k}} = \underline{\mathbf{S}}^{\mathbf{k}}$. In practice, this entails that excitation energies will be obtained here in the Franck-Condon approximation. [30]

Substituting the matrix expansions of Eqn. (17) into Eq. (9) and neglecting all resulting terms that are of order 2 or higher in $\lambda^{a}_{\pm\omega}$ gives the pair of equations

$$\mathbf{F}^{\mathbf{k},a}_{\pm\omega}\underline{\mathbf{C}}^{\mathbf{k}} + \underline{\mathbf{F}}^{\mathbf{k}}\mathbf{C}^{\mathbf{k},a}_{\pm\omega} \pm \omega\mathbf{C}^{\mathbf{k},a}_{\pm\omega} = \underline{\mathbf{S}}^{\mathbf{k}}\mathbf{C}^{\mathbf{k},a}_{\pm\omega}\underline{\mathbf{E}}^{\mathbf{k}} + \underline{\mathbf{S}}^{\mathbf{k}}\underline{\mathbf{C}}^{\mathbf{k}}\mathbf{E}^{\mathbf{k},a}_{\pm\omega} \tag{20}$$

for $+\omega$ and $-\omega$. These equations can be converted to CO representation by multiplying on the left by $\underline{\mathbf{C}}^{\mathbf{k}\dagger}$. Using the definitions

$$\mathbf{1} = \underline{\mathbf{C}}^{\mathbf{k}\dagger}\underline{\mathbf{S}}^{\mathbf{k}}\underline{\mathbf{C}}^{\mathbf{k}} \tag{21}$$

$$\tilde{\mathbf{C}}^{\mathbf{k},a}_{\pm\omega} = \underline{\mathbf{C}}^{\mathbf{k}}\mathbf{U}^{\mathbf{k},a}_{\pm\omega} \tag{22}$$

$$\tilde{\mathbf{G}}^{\mathbf{k},a}_{\pm\omega} = \underline{\mathbf{C}}^{\mathbf{k}\dagger}\mathbf{F}^{\mathbf{k},a}_{\pm\omega}\underline{\mathbf{C}}^{\mathbf{k}} \tag{23}$$

for the expressions resulting from Eqn. (20) gives the frequency-dependent coupled-perturbed equations

$$\tilde{\mathbf{U}}^{\mathbf{k},a}_{\pm\omega}\underline{\tilde{\mathbf{E}}}^{\mathbf{k}} - \underline{\tilde{\mathbf{E}}}^{\mathbf{k}}\tilde{\mathbf{U}}^{\mathbf{k},a}_{\mp\omega} \pm \omega\tilde{\mathbf{U}}^{\mathbf{k},a}_{\pm\omega} = \tilde{\mathbf{G}}^{\mathbf{k},a}_{\pm\omega} - \tilde{\mathbf{E}}^{\mathbf{k},a}_{\pm\omega}. \tag{24}$$

A tilde indicates that the corresponding matrix is in CO representation. The matrix $\mathbf{U}^{\mathbf{k},a}_{\pm\omega}$ introduced in Eqn. (22) has to be interpreted as a unitary matrix that transforms the unperturbed CO coefficients into their linear response by linearly combining occupied and unoccupied one-particle states of the unperturbed Hamiltonian. As will be shown below, various frequency-dependent response functions can directly be determined from $\mathbf{U}^{\mathbf{k},a}_{\pm\omega}$. The RPA-matrix formalism, Eqn. (3), can easily be derived from Eqns. (24). [25] This will allow us to identify the poles of response functions computed *via* Eqns. (24) corresponding to the *optically allowed* (forced) resonances described by Eqn. (3) with fictitious transitions between stationary electronically excited states. [16]

The matrix $\tilde{\mathbf{G}}^{\mathbf{k},a}_{\pm\omega}$ is the sum of the response of the Fock/Kohn-Sham matrix, $\mathbf{B}^{\mathbf{k},a}_{\pm\omega}$, and of the matrix representation of the external perturbation $\boldsymbol{\Omega}^{\mathbf{k},a}$ from Eqn. (15), *i.e.*

$$\tilde{\mathbf{G}}^{\mathbf{k},a}_{\pm\omega} = \underline{\mathbf{C}}^{\mathbf{k}\dagger}\mathbf{B}^{\mathbf{k},a}_{\pm\omega}\underline{\mathbf{C}}^{\mathbf{k}} + \underline{\mathbf{C}}^{\mathbf{k}\dagger}\boldsymbol{\Omega}^{\mathbf{k},a}\underline{\mathbf{C}}^{\mathbf{k}}. \tag{25}$$

The response matrix $\mathbf{B}^{\mathbf{k},a}_{\pm\omega}$ is related to the linear density matrix response $\mathbf{D}^{\mathbf{k},a}_{\pm\omega}$ by

$$B^{\mathbf{k},a}_{\pm\omega,\mu\nu} = \sum_{\rho\tau}^{N} D^{\mathbf{k},a}_{\pm\omega,\rho\tau}(\mu\nu||\rho\tau) \tag{26}$$

where we have used the standard notation for the two-electron integrals, which are defined:
1) for Hartree-Fock:

$$\begin{aligned}(\mu\nu||\rho\tau)^{\mathrm{HF}} &= (\mu\nu|\rho\tau) - (\mu\rho|\nu\tau) \\ &= \int d\mathbf{r}d\mathbf{r}' \frac{\chi_\mu^*(\mathbf{r})\chi_\nu(\mathbf{r})\chi_\rho^*(\mathbf{r}')\chi_\tau(\mathbf{r}')}{|\mathbf{r}-\mathbf{r}'|} - \int d\mathbf{r}d\mathbf{r}' \frac{\chi_\mu^*(\mathbf{r})\chi_\rho(\mathbf{r})\chi_\nu^*(\mathbf{r}')\chi_\tau(\mathbf{r}')}{|\mathbf{r}-\mathbf{r}'|},\end{aligned} \tag{27}$$

2) for pure DFT:

$$\begin{aligned}(\mu\nu||\rho\tau)^{\mathrm{KS}} &= (\mu\nu|\rho\tau) + (\mu\nu|f_{\mathrm{xc}}|\rho\tau) \\ &= \int d\mathbf{r}d\mathbf{r}' \frac{\chi_\mu^*(\mathbf{r})\chi_\nu(\mathbf{r})\chi_\tau^*(\mathbf{r}')\chi_\rho(\mathbf{r}')}{|\mathbf{r}-\mathbf{r}'|} + \int d\mathbf{r}d\mathbf{r}' \chi_\mu^*(\mathbf{r})\chi_\nu(\mathbf{r}) f_{\mathrm{xc}} \chi_\rho^*(\mathbf{r}')\chi_\tau(\mathbf{r}'),\end{aligned} \tag{28}$$

3) for hybrid Hartree-Fock/DFT:

$$(\mu\nu||\rho\tau)^{\mathrm{hybrid}} = (\mu\nu|\rho\tau) + (1 - c_{\mathrm{HF}})(\mu\nu|f_{\mathrm{xc}}|\rho\tau) - c_{\mathrm{HF}}(\mu\rho|\nu\tau). \tag{29}$$

The linear density matrix response in Eqn. (26) can be computed from the matrix $\mathbf{U}^{\mathbf{k},a}_{\pm\omega}$ appearing in Eqn. (24):

$$D^{\mathbf{k},a}_{\pm\omega,\mu\nu} = \sum_i^{N_{\mathrm{occ}}} \sum_d^{N_{\mathrm{vir}}} \sum_{\rho\tau}^{N} ([\tilde{U}^{\mathbf{k},a}_{\pm\omega}]^*_{id} \underline{C}^{\mathbf{k}*}_{\mu d} \underline{C}^{\mathbf{k}}_{\nu i} + \underline{C}^{\mathbf{k}*}_{\mu i} [\tilde{U}^{\mathbf{k},a}_{\pm\omega}]_{id} \underline{C}^{\mathbf{k}}_{\nu d}) \tag{30}$$

Here N_{occ} and N_{vir} are the number of occupied and unoccupied COs respectively, and $N = N_{\mathrm{occ}} + N_{\mathrm{vir}}$. Finally, the matrix of the external perturbation $\mathbf{\Omega}^{\mathbf{k},a}$ in the right hand side of Eqn. (25) is the atomic orbital representation of the the operator $\hat{\mathbf{r}}$ of Eqn. (16), *i.e.*

$$\Omega^{\mathbf{k},a}_{\mu\nu} = i\langle\phi_\mu(\mathbf{r},\mathbf{k})|e^{i\mathbf{k}\cdot\mathbf{r}}\nabla_{k_a}e^{-i\mathbf{k}\cdot\mathbf{r}}|\phi_\nu(\mathbf{r},\mathbf{k})\rangle. \tag{31}$$

2.2. *Self-consistent solution of the coupled-perturbed equations*

The coupled-perturbed equations, Eqns. (24), can be solved symbolically to give

$$\tilde{U}^{\mathbf{k},a}_{\pm\omega,id} = \frac{\tilde{G}^{\mathbf{k},a}_{\pm\omega,id}}{\varepsilon^{\mathbf{k}}_d - \varepsilon^{\mathbf{k}}_i \mp \omega} = \frac{[\underline{\mathbf{C}}^{\mathbf{k}\dagger}(\mathbf{\Omega}^{\mathbf{k},a} + \mathbf{B}^{\mathbf{k},a}_{\pm\omega})\underline{\mathbf{C}}^{\mathbf{k}}]_{id}}{\varepsilon^{\mathbf{k}}_d - \varepsilon^{\mathbf{k}}_i \mp \omega}, \tag{32}$$

where $\varepsilon^{\mathbf{k}}_i$ and $\varepsilon^{\mathbf{k}}_d$ are matrix elements of $\underline{\mathbf{E}}^{\mathbf{k}}$ in Eqn. (18), *i.e.* unperturbed one-particle energies at $\mathbf{k}$ corresponding to occupied (i) and unoccupied (d) states. Since, from Eqn. (26), $\mathbf{B}^{\mathbf{k},a}_{\pm\omega}$ depends explicitly on $\tilde{\mathbf{U}}^{\mathbf{k},a}_{\pm\omega}$, Eqn. (32) defines an infinite recursion, which can be solved *e.g.* using diagrammatic techniques (see *e.g.* Ref. [31]). Here we adopt a different approach, and we recast Eqn. (32) as a self-consistent procedure, exploiting the ground-state self-consistent field solver implemented in the `CRYSTAL` code. [21] Considering a pair of *fixed* frequencies $\pm\omega = \pm\omega_I$, we start by computing a zeroth order matrix $\tilde{\mathbf{U}}^{\mathbf{k},a}_{\pm\omega}$ in the absence of inter-electron contributions by setting $\mathbf{B}^{\mathbf{k},a}_{\pm\omega} = \mathbf{0}$ in Eqn. (32). In Hartree-Fock approximation, this iteration corresponds to the so-called "uncoupled Hartree-Fock" perturbation theory. [31] From the resulting $\tilde{\mathbf{U}}^{\mathbf{k},a}_{\pm\omega}$ we compute the density matrix response $\mathbf{D}^{\mathbf{k},a}_{\pm\omega}$ using Eqn. (30), and this allows us now to estimate the two-electron response matrix from Eqn. (26). We can then in turn compute the first-iteration

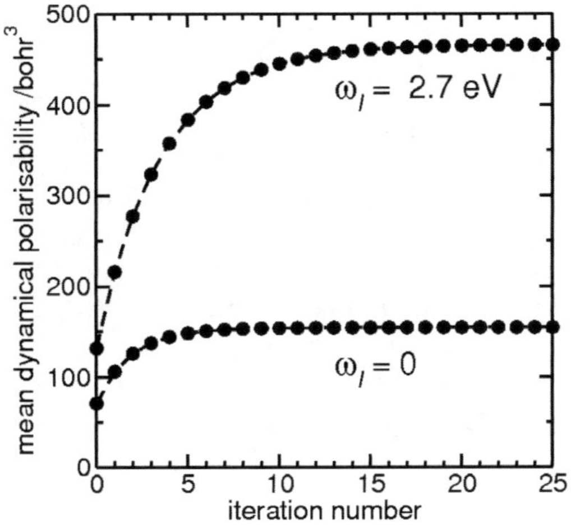

Figure 1. Convergence of coupled mean dynamical polarisability for crystalline silicon with TD-HF at $\omega_I = 0$ and 2.7 eV during the self-consistent solution of Eqn. (32).

coupled-perturbed matrix $\tilde{\mathbf{U}}^{\mathbf{k},a}_{\pm\omega}$, and we then repeat the same procedure for all subsequent iterations. At each iteration we estimate a (coupled) polarisability tensor, [20, 25, 32, 33]

$$\alpha_{ab}(\pm\omega_I) = -2\sum_{\mathbf{k}}^{N_\mathbf{k}} w_\mathbf{k} \sum_i^{N_{\rm occ}} \sum_d^{N_{\rm vir}} [\tilde{U}^{\mathbf{k},a}_{\pm\omega_I}]_{di}\tilde{\Omega}^{\mathbf{k},b}_{id} + \tilde{\Omega}^{\mathbf{k},b}_{di}[\tilde{U}^{\mathbf{k},a}_{\pm\omega_I}]_{id}, \tag{33}$$

where a and b indicate Cartesian components, $N_\mathbf{k}$ is the number of $\mathbf{k}$-points in the Brillouin zone, with $w_\mathbf{k}$ the weight of $\mathbf{k}$ in Brillouin zone integrals. $\tilde{\Omega}^{\mathbf{k},b}_{id}$ is the CO representation of Eqn. (31),

$$\tilde{\Omega}^{\mathbf{k},b}_{id} = i\langle\psi_i(\mathbf{r},\mathbf{k})|e^{i\mathbf{k}\cdot\mathbf{r}}\nabla_{k_b}e^{-i\mathbf{k}\cdot\mathbf{r}}|\psi_d(\mathbf{r},\mathbf{k})\rangle, \tag{34}$$

and contains the Cartesian component b of the position operator matrix elements between independent-particle states i and d at $\mathbf{k}$. We then use the mean dynamical polarisability

$$\overline{\alpha}(\pm\omega_I) = \frac{1}{3}\mathrm{tr}\{\boldsymbol{\alpha}(\pm\omega_I)\}, \tag{35}$$

to define a convergence criterion for self-consistency in Eqn. (32), see Figure 1. From the self-consistent polarisability, the frequency dependent dielectric tensor $\boldsymbol{\epsilon}(\omega)$, the first-order dielectric susceptibility $\boldsymbol{\chi}^{(1)}(\omega)$ and the (complex) refractive index $\mathbf{n}(\omega)$ can also be computed using the relations [21]

$$\epsilon_{ab}(\pm\omega_I) = n^2_{ab}(\pm\omega_I) = 1 + \chi^{(1)}_{ab}(\pm\omega_I) = 1 + \frac{4\pi}{V}\alpha_{ab}(\pm\omega_I), \tag{36}$$

where V is the unit cell volume.

2.3. Excitation energies

At iteration 0, the uncoupled matrix $\tilde{\mathbf{U}}^{\mathbf{k},a}_{\pm\omega}$ at $\omega = \omega_I$ is given by

$$\tilde{U}^{\mathbf{k},a}_{\omega_I,id} = \frac{\tilde{\Omega}^{\mathbf{k},a}_{id}}{\varepsilon^\mathbf{k}_d - \varepsilon^\mathbf{k}_i - \omega_I}, \tag{37}$$

where Eqns. (32) and (34) have been used. It is clear from this expression that $\tilde{U}^{\mathbf{k},a}_{\omega_I,id}$ diverges whenever ω_I approaches an independent-particle energy difference $\varepsilon^\mathbf{k}_d - \varepsilon^\mathbf{k}_i$, provided

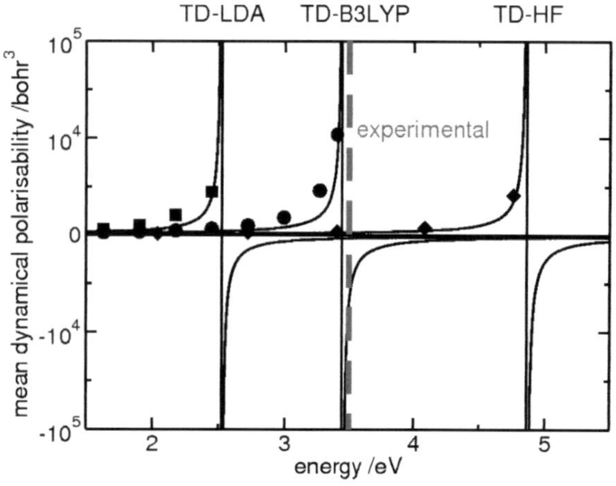

Figure 2. Energy dependence of the mean dynamical polarisability of crystalline silicon with TD-LDA, TD-B3LYP and TD-HF, showing the position of the pole corresponding to the lowest optically-allowed many-body excitation. The dashed vertical line is the experimental value of the optical gap.

the corresponding independent-particle free oscillation couples to the external field, *i.e.* $\tilde{\Omega}_{id}^{\mathbf{k},a} \neq 0$. Because of Eqn. (33), independent-particle dipole allowed electronic transitions therefore correspond to the poles of the polarisability tensor $\boldsymbol{\alpha}(\omega)$, or equivalently of the mean dynamical polarisability $\overline{\alpha}(\pm\omega)$, with non-zero residues.

For the fully coupled self-consistent matrix $\tilde{\mathbf{U}}_{\pm\omega}^{\mathbf{k},a}$ solving Eqn. (32), we can identify the poles of $\overline{\alpha}(\pm\omega)$ with the *many-body* excitation modes that couple with the external field (see Section 2.1). Therefore, similar to the method of Eqn. (1), Eqn. (32) determines the forced oscillations induced by the external field in the interacting electron system. Many-body excitation energies can then straightforwardly be computed by solving Eqn. (32) over a given frequency range. Adaptive algorithms based on the on-the-fly estimation of the the derivative of $\overline{\alpha}(\pm\omega)$ with respect to ω can be devised to locate the frequency of the poles with an accuracy at least comparable to the iterative (*e.g.* Davidson-like [34, 35]) diagonalisation methods typically used to solve large eigenvalue problems like Eqn. (3).

It is interesting to observe that the calculation of the response matrix $\tilde{\mathbf{U}}_{\pm\omega}^{\mathbf{k},a}$ in Eqn. (32) and of the polarisability in Eqn. (33) only involve *vertical* transitions between the occupied and unoccupied one-particle states at each $\mathbf{k}$ point. This is a consequence of the $\mathbf{k}$ factorisability of the linear density matrix response $D_{\pm\omega,\rho\tau}^{\mathbf{k},a}$, and therefore of the two-electron contribution to the Kohn-Sham/Fock matrix response $B_{\pm\omega,\mu\nu}^{\mathbf{k},a}$, *cf.* Eqn. (26). The matrix $\tilde{\boldsymbol{\Omega}}^{\mathbf{k},b} = \underline{\mathbf{C}}^{\mathbf{k}\dagger}\boldsymbol{\Omega}^{\mathbf{k},b}\underline{\mathbf{C}}^{\mathbf{k}}$ in the right hand side of Eqn. (32) is also factorisable, owing to the fact that matrix elements of $\nabla_{\mathbf{k}}$ between Bloch orbitals,

$$\langle \mathbf{k}'m|\nabla_{\mathbf{k}}|\mathbf{k}n\rangle = \langle u_{\mathbf{k}'m}|e^{i(\mathbf{k}-\mathbf{k}')\cdot\mathbf{r}}\nabla_{\mathbf{k}}u_{\mathbf{k}n}\rangle, \tag{38}$$

are zero unless $\mathbf{k}' = \mathbf{k}$, since $u_{\mathbf{k}'m}^{*}(\mathbf{r})\nabla_{\mathbf{k}}u_{\mathbf{k}n}(\mathbf{r})$ is lattice periodic. [27, 28] The scaling of the self-consistent procedure used to solve Eqn. (32) is therefore linear in the number of $\mathbf{k}$ points.

3. Results and discussion

We show in Figure 2 the dependence of the coupled mean dynamical polarisability on the absorption energy in crystalline silicon, computed by including only the lowest optical excitation. Details of the calculations are as from Ref. [25]. In these examples, the energy of the lowest allowed excitation, corresponding to the *optical* gap of the system, was determined by solving Eqn. (32) at several values of ω_I (filled shapes in Figure 2) and interpolating the results with an inverse law, $\overline{\alpha}(\omega_i) = [A + B(\omega_i - C)]^{-1}$ (continuous lines). We remark that the profile of the mean dynamical polarisability extrapolated for $\omega > \omega_I$ assumes that only one many-body excitation at ω_I contributes to the absorption process, and does not therefore necessarily

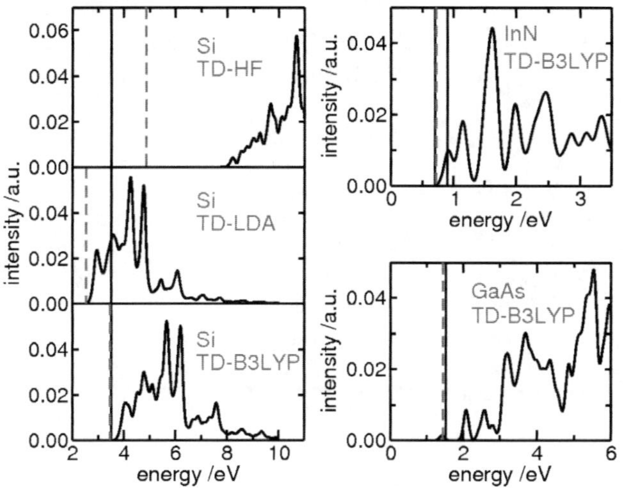

Figure 3. (*Left*) Independent-particle absorption spectra of crystalline silicon at the Hartree-Fock, LDA and B3LYP levels of theory compared to the optical gaps obtained from TD-HF, TD-LDA and TD-B3LYP (dashed lines). Continuous vertical lines are experimental estimates of the optical gaps. (*Right*) Independent-particle spectra and optical gaps of InN and GaAs from B3LYP/TD-B3LYP.

correspond to a physical situation. The three approximations used here, TD-HF, TD-LDA, and TD-B3LYP use the two-electron integrals of Eqns.(27), (28) and (29) respectively in Eqn. (26). For the exchange-correlation kernel $f_{\rm xc}$ in Eqns. (28) and (29) we used the adiabatic local density approximation [12] (ALDA). The calculated values of the optical gaps are $E_o^{\rm HF} = 4.86$ eV, $E_o^{\rm LDA} = 2.52$ eV, $E_o^{\rm B3LYP} = 3.44$ eV, whereas the experimental estimate is $E_o = 3.5$ eV. [36, 37]

In Figure (3) (left) we compare the calculated optical gaps for silicon with the corresponding independent-particle absorption spectra,

$$I(\omega) = \sum_{\mathbf{k}}^{N_{\mathbf{k}}} w_{\mathbf{k}} \sum_{i}^{N_{\rm occ}} \sum_{d}^{N_{\rm vir}} \tilde{\Omega}_{id}^{\mathbf{k},b}\ \delta(\varepsilon_d^{\mathbf{k}} - \varepsilon_i^{\mathbf{k}} - \omega). \tag{39}$$

TD-HF yields a large negative shift in the optical gap relative to the lowest energy Hartree-Fock absorption (~ 2.5 eV), but the resulting optical gap remains far too large compared to the experimental value. The huge overestimation of the independent-particle gap is a consequence of the virtual Hartree-Fock states lying at too high energies, owing to the fact that an electron promoted to an empty state experiences repulsion with N (rather than $N-1$) electrons. [15] The large negative shift in absorption energy brought about by TD-HF is related to the overestimation of the Coulomb attraction between the excited electron and the hole, which are only partially screened by many-body correlation effects. Overall, the TD-HF correction is however insufficient to correct the unphysically high virtual Hartree-Fock orbital energies. In LDA, the onset of the independent-particle absorption occurs at far too low energies. This is a consequence of the underestimation of the band gap in LDA. The TD-LDA correction, in the ALDA approximation, appears to be modest. Consistent with a number of studies on the performance of the ALDA kernel for extended systems in TD-DFT, we find here that the optical gap essentially occurs at the same energy as the lowest independent-particle LDA excitation, and both are largely underestimated relative to experiment. Finally, TD-B3LYP yields a moderate stabilisation of the particle-hole pair (~ 0.4 eV) relative to the independent-particle gap, [38, 39] and the resulting optical gap is in close agreement with experiment. This approach is therefore capable of reproducing the formation of bound excitonic pairs in silicon. [40]

TD-B3LYP yields accuracy comparable to the case of crystalline silicon (≤ 0.1 eV difference from experiment) for a number of other weakly bound excitonic semiconductors (Table 1), including technologically important direct-gap III-V semiconductors binaries. Satisfactory agreement with experiment is also observed in (at least some classes of) tightly bound excitons in

Journal of Physics: Conference Series **367** (2012) 012001 doi:10.1088/1742-6596/367/1/012001

Table 1. Calculated optical gaps compared to experimental values. n/a indicates the optical gap is lower than 0.1 eV. Experimental data is from Refs. [36, 37] for all systems except SiO_2 (Ref. [41]) and LiF (Ref. [42]).

	TD-HF	TD-LDA	TD-B3LYP	Expt.
Si	4.86	2.52	3.44	3.5
GaAs	4.94	0.21	1.44	1.52
InSb	3.15	n/a	0.25	0.24
GaN	8.83	1.91	3.52	3.51
InN	4.95	n/a	0.72	0.7-0.9
MgO	13.61	4.87	7.02	7.16
ZnO	8.65	0.95	3.02	3.4
SiO_2	-	-	9.99	10.3
LiF	16.3	-	11.90	12.8

insulators, like SiO_2, although self-trapping of the excitons can complicate the direct comparison with experiment in many cases. [43, 44] The ability of TD-B3LYP to describe accurately bound excitons in these cases is likely to be explained by the proper inclusion of the non-local interaction between an excited electrons and a hole, which results in the appearance of an electron-hole Coulomb-like attraction term, $f_{\text{eh}}(\mathbf{r}, \mathbf{r}') \propto -c_{\text{HF}}/|\mathbf{r}-\mathbf{r}'|$, in the response equations. This fact has been exploited in the construction of semi-empirical adiabatic response kernels that can account very accurately for the optical reponse of weakly bound excitonic systems. [2, 3, 40] In our formalism, the electron-hole Coulomb attraction term appears in consequence to the inclusion of the Fock exchange integrals, $-(\mu\rho|\nu\tau)$, in Eqn. (29). [45, 46] It is interesting to observe that our approach, based on the self-consistent determination of the Kohn-Sham linear density *matrix* response (rather than solving directly for the linear density response, as is common in implementations of TD-DFT based on variational density-functional perturbation theory [29, 47–53]), treats the Fock exchange contribution, which can only be formulated as a non-adiabatic functional of the density, [16] as a *purely adiabatic* functional of the Kohn-Sham density matrix.

More complex appears to be the case of charge-transfer (CT) excitons, see *e.g.* LiF in Table 1, in which the electron and the hole localise in disconnected and possibly remote regions of space, or on different crystal sublattices. Similar to CT excitations in gas-phase molecules, [45, 54] TD-DFT with local, semi-local or hybrid XC functionals and adiabatic XC kernels yields CT energies that are far too low compared to experiment in the condensed phase. [29, 46, 55, 56] This failure may be interpreted in terms of the incorrect asymtpotics of f_{eh}, which, for $|\mathbf{r}-\mathbf{r}'| \to \infty$, does not reproduce the proper Coulomb tail $-1/|\mathbf{r}-\mathbf{r}'|$. Various methods have been proposed in the quantum-chemical literature to overcome this well-known limitation of (adiabatic) TD-DFT, typically based on imposing the correct asymptotics in the electron-hole attraction, *e.g.* using range-separated functionals in which the fraction of Fock exchange varies with the distance between particle and hole, [57–61] or *via* model response kernels describing the physical factors contributing to the CT process. [62–64] Our method can be extended to encompass these classes of approaches, which only require modifying the algorithm at the level of the calculation of the two-electron integrals, Eqns. (27)-(29). A more elegant and potentially much more far reaching approach that has been proved to describe accurately wide classes of excitons (including, in particular, the CT exciton of LiF) has recently been proposed by Sharma *et al.*, [65] based on a *bootstrap* XC kernel that is computed self-consistently from the dielectric function during the

Journal of Physics: Conference Series **367** (2012) 012001 doi:10.1088/1742-6596/367/1/012001

solution of the exact Dyson equation for the response.

4. Summary and conclusions

We have described a method for computing the electronic response of extended (crystalline) systems that accounts accurately and computationally efficiently for optical excitations, including those involving interactions between excited electrons and holes. Our approach can be implemented at various levels of theory (pure TD-DFT, TD-HF, hybrid TD-DFT) and shows a very favourable scaling with system size and/or number of **k** points in Brillouin zone integrals. In its hybrid TD-DFT implementation, it achieves predictive accuracy for weakly bound excitonic semiconductors, without resorting to semi-empirical response kernels and/or *a posteriori* quasi-particle corrections of one-particle Kohn-Sham energies. Good agreement with experiment is also observed for tightly-bound excitons in insulators, whereas CT exciton energies are largely underestimated, similar to CT excitations in molecules. Possible generalisations to address these classes of excitations are currently being explored.

References

[1] Gross E K U and Kohn W 1990 *Adv. Quant. Chem.* **21** 255
[2] Onida G, Reining L and Rubio A 2002 *Rev. Mod. Phys.* **74** 601
[3] Botti S, Schindlmayr A, Sole R D and Reining L 2007 *Rep. Prog. Phys.* **70** 357
[4] Burke K, Werschnik J and Gross E K U 2005 *J. Chem. Phys.* **123** 062206
[5] Casida M E 2009 *J. Mol. Struct. (THEOCHEM)* **914** 3
[6] Ullrich C 2012 *Time-Dependent Density-Functional Theory: Concepts and Applications* (Oxford University Press)
[7] Runge E and Gross E K U 1983 *Phys. Rev. Lett.* **52** 997
[8] Petersilka M, Gossmann U J and Gross E K U 1996 *Phys. Rev. Lett.* **76** 1212
[9] Petersilka M, Gross E K U and Burke K 2000 *Int. J. Quant. Chem.* **80** 534
[10] Hohenberg P and Kohn W 1964 *Phys. Rev.* **136** B864
[11] Kohn W and Sham L J 1965 *Phys. Rev.* **140** A1133
[12] Zangwill A and Soven P 1981 *Phys. Rev. B* **24** 4121
[13] Casida M E, Ipatov A and Cordova F 2006 *Time-Dependent Density-Functional Theory* ed Marques M A L, Ullrich C, Nogueira F, Rubion A and Gross E K U (Springer, Berlin)
[14] Dreuw A and Head-Gordon M 2005 *Chem. Rev.* **105** 4009
[15] Szabo A and Ostlund N S 1996 *Modern Quantum Chemistry* (Dover Publications)
[16] McLachlan A D and Ball M A 1964 *Reviews of Modern Physics* **36** 844
[17] Becke A D 1993 *J. Chem. Phys.* **98** 1372
[18] Hirata S, Head-Gordon M and Bartlett R J 1999 *J. Chem. Phys.* **111** 10774
[19] Hurst G J B and Dupuis M 1988 *J. Chem. Phys.* **89** 385
[20] Kirtman B, Gu F L and Bishop D M 2000 *J. Chem. Phys.* **113** 1294
[21] Ferrero M, Rerat M, Orlando R and Dovesi R 2008 *J. Comput. Chem.* **29** 1450
[22] Dovesi R, Orlando R, Civalleri B, Roetti R, Saunders V R and Zicovich-Wilson C M 2005 *Z. Kristallogr.* **220** 571
[23] Dovesi R *et al.*, CRYSTAL09 User's manual; http://www.crystal.unito.it/
[24] Bush I J, Tomić S, Searle B G, Mallia G, Bailey C L, Montanari B, Bernasconi L, Carr J M and Harrison N M 2011 *Proc. Roy. Soc. A* **467** 2112
[25] Bernasconi L, Tomić S, Ferrero M, Rérat M, Orlando R, Dovesi R and Harrison N M 2011 *Phys. Rev. B* **83** 195325
[26] Sekino H and Bartlett R J 1986 *J. Chem. Phys* **85** 976
[27] Blount E I 1962 *Solid State Physics* (Academic, New York)
[28] Otto P 1992 *Phys. Rev. B* **45** 10876
[29] Bernasconi L, Sprik M and Hutter J 2003 *J. Chem. Phys.* **119** 12417
[30] Atkins P W and Friedman R S 1999 *Molecular Quantum Mechanics* (Oxford University Press)
[31] Caves T C and Karplus M 1969 *J. Chem. Phys.* **50** 3649
[32] Bishop D M and Kirtman B 1991 *J. Chem. Phys.* **95** 2646
[33] Ferrero M, Rérat M, Orlando R and Dovesi R 2008 *J. Chem. Phys.* **128** 014110
[34] Davidson E R 1975 *Journal of Computational Physics* **17** 87
[35] Stratmann R E, Scuseria G E and Frisch M J 1998 *J. Chem. Phys.* **109** 8218

[36] Madelung O 2004 *Semiconductors: Data Handbook* (Springer, Berlin)
[37] Heyd J, Peralta J E, Scuseria G E and Martin R L 2005 *J. Chem. Phys.* **123** 174101
[38] Muscat J, Wander A and Harrison N M 2001 *Chem. Phys. Lett.* **342** 397
[39] Tomić S, Montanari B and Harrison N M 2008 *Physica E* **40** 2125
[40] Reining L, Olevano V, Rubio A and Onida G 2002 *Phys. Rev. Lett.* **88** 066404
[41] Ramos L E, Furthmuller J and Bechstedt F 2004 *Phys. Rev. B* **69** 085102
[42] Roessler D M and Walker W C 1967 *J. Phys. Chem. Solids* **28** 1507
[43] Trukhin A N 1992 *Journal of Non-Crystalline Solids* **149** 32
[44] Ismail-Beigi S and Louie S G 2005 *Phys. Rev. Lett.* **95** 156401
[45] Dreuw A, Weisman J L and Head-Gordon M 2003 *J. Chem. Phys.* **119** 2943
[46] Bernasconi L, Sprik M and Hutter J 2004 *Chem. Phys. Lett.* **394** 141
[47] Gonze X 1995 *Phys. Rev. A* **52** 1096
[48] Filippi C, Umrigar C J and Gonze X 1997 *J. Chem. Phys.* **107** 9994
[49] Putrino A, Sebastiani D and Parrinello M 2000 *J. Chem. Phys.* **113** 7102
[50] Baroni S, de Gironcoli S, Corso A D and Giannozzi P 2001 *Rev. Mod. Phys.* **73** 515
[51] Hutter J 2003 *J. Chem. Phys.* **118** 3928
[52] Walker B, Saitta A M, Gebauer R and Baroni S 2006 *Phys. Rev. Lett.* **113001** 113001
[53] Rocca D, Gebauer R, Saad Y and Baroni S 2008 *J. Chem. Phys.* **128** 154105
[54] Tozer D J, Amos R D, Handy N C, Roos B O and Serrano-Andrés L 1999 *Mol. Phys.* **97** 859
[55] Bernasconi L, Blumberger J, Sprik M and Vuilleumier R 2004 *J. Chem. Phys.* **121** 11885
[56] Bernasconi L 2010 *J. Chem. Phys.* **132** 184513
[57] Chiba M, Tsuneda T and Hirao K 2006 *J. Chem. Phys.* **124** 144106
[58] Tawada Y, Yanagisawa T T S, Yanai T and Hirao K 2004 *J. Chem. Phys.* **120** 8425
[59] Peach M J G, Helgaker T, Salek P, Keal T W, Lutnaes O B, Tozer D J and Handy N C 2006 *PhysChemChemPhys* **8** 558
[60] Yanai T, Tew D P and Handy N C 2004 *Chem. Phys. Lett.* **393** 51
[61] Rudberg E, Salek P, Helgaker T and Agren H 2005 *J. Chem. Phys.* **123** 184108
[62] Casida M E, Gutierrez F, Guan J, Gadea F X, Salahub D and Daudey J P 2000 *J. Chem. Phys.* **113** 7062
[63] Gritsenko O and Baerends E J 2004 *J. Chem. Phys.* **121** 655
[64] Neugebauer J, Gritsenko O and Baerends E J 2006 *J. Chem. Phys.* **124** 214102
[65] Sharma S, Dewhurst J K, Sanna A and Gross E K U 2011 *Phys. Rev. Lett.* **107** 186401

Linear-Scaling Density Functional Theory Simulations of Polar Semiconductor Nanorods

Nicholas D. M. Hine, Philip W. Avraam, Paul Tangney and Peter D. Haynes

Department of Physics and Department of Materials, Imperial College London, Exhibition Road, London SW7 2AZ, United Kingdom

E-mail: nicholas.hine@imperial.ac.uk

Abstract. Binary polar semiconductors in the wurtzite structure can be grown in the form of nanorods aligned along the wurtzite [0001] direction. In such structures, very large dipole moments have been observed experimentally. We have studied the distribution of charge in GaAs nanorods, so as to elucidate the origin of the dipole moments. To make contact with the realistic experimental regime, we need to model systems of several thousand atoms, necessitating the use of a Linear-Scaling formulation of DFT. We use the ONETEP code, which combines the benefits of $O(N)$ computational effort and plane-wave accuracy. We show that both the direction and magnitude of the dipole moment of a nanorod, and its electric field, depend sensitively on how its surfaces are terminated and do not depend strongly on the spontaneous polarization of the underlying lattice. Furthermore, we observe that the Fermi level for an isolated nanorod always coincides with a significant density of electronic surface states at the polar surfaces, which are either mid-gap states or band-edge states. These states pin the Fermi level, and therefore fix the potential difference across the rod. We provide evidence that this effect can have a determining influence on the polarity of nanorods, and has consequences for the way a rod responds to changes in its surface chemistry.

1. Introduction

Nanostructured materials are becoming ever more important due to the possibilities they enable for control and optimisation of materials properties. Within this broad class of materials, semiconductor nanocrystals have a prominent role, with current and future applications in optoelectronics, thermoelectrics, and photovoltaics, and as pressure sensors and catalysts. The valuable electronic and optical properties of semiconductor nanocrystals are strongly connected to their size and shape [1, 2, 3, 4, 5, 6, 7, 8]. Nanocrystals of binary semiconductors, particularly, though not exclusively, those with noncentrosymmetric crystal structures such as wurtzite, have been shown experimentally to exhibit large dipole moments [9, 10, 11]. These dipole moments strongly affect their internal electronic structure and therefore their optical properties, as well as their interactions with the environment and with other nanocrystals. However, the origin of these dipole moments, and indeed the factors influencing the charge distribution generally, are not well-understood.

Developments in electronic structure theory, particularly the availability of accurate quantum-mechanical simulation methods with linear-scaling computational cost [12, 13, 14, 15, 16, 17, 18, 19, 20, 21], make it possible to simulate whole nanostructures of realistic size. This raises the possibility of simulating realistic nanoparticles and systematically varying controllable aspects such as sizes, shapes, compositions and surface terminations. The development of linear-scaling DFT methodologies which combine low-order-scaling with systematic, plane-wave accuracy, as in the ONETEP code [21, 22], in particular, make it possible to disentangle the competing factors which determine the distribution of charge in such systems.

We have performed LS-DFT calculations [23], using the ONETEP code, of the ground-state charge distributions in wurtzite-structure GaAs nanorods of sizes comparable to those found in experiment. We showed that the dipole moment observed is strongly dependent on the surface termination. In particular, there is a correlation between the dipole moment and the energy gap observed between the position of the highest occupied state at one polar surface and the lowest unoccupied state at the other polar surface.

Subsequently [24], we extended these simulations by examining the behaviour of the dipole moment as a function of the length and cross-sectional area of the rod, and the composition of the rod as it varies between different III-V element pairs. We introduced an interpretation of these behaviours in terms of the concept of *Fermi level pinning* (FLP). FLP is known to occur in semi-infinite semiconductor surfaces when there are partially-occupied states at the Fermi level, and can be expected to occur when the dipole moment of a rod is large enough that the resulting electric field adjusts the Fermi level enough that the potential difference along the rod is comparable to the bandgap. We showed that a finite-surface version of FLP then plays a crucial role in determining the polar characteristics of nanorod systems.

In the current article, we present two new features: firstly, we present a detailed description of the methodology of these simulations (Section 2), which could not be included in the previous works due to limitations of space, and secondly, we revisit the results of our earlier work [23] on variation of dipole moment with surface termination, to examine how they can be interpreted in terms of the concept of Fermi level pinning (Section 3).

2. Simulation Methodology

2.1. Electronic Energy Minimisation

This work uses linear-scaling density-functional theory (LS-DFT) as implemented in the ONETEP code [21, 22]. We represent the single-particle density matrix, which contains a complete description of the Kohn-Sham system, using a set of non-orthogonal, strictly localized orbitals $\{\phi_\alpha(\mathbf{r})\}$, as

$$\rho(\mathbf{r},\mathbf{r}') = \sum_{\alpha\beta} \phi_\alpha(\mathbf{r}) K^{\alpha\beta} \phi^*_\beta(\mathbf{r}'), \qquad (1)$$

where the density kernel $K^{\alpha\beta}$ is a generalisation of occupancies to a nonorthogonal representation. The orbitals $\{\phi_\alpha(\mathbf{r})\}$ are referred to as Non-orthogonal Generalised Wannier Functions (NGWFs). Each of these NGWFs is expressed using an underlying basis of periodic cardinal sine, or 'psinc' functions, $D_i(\mathbf{r})$, which are effectively representations of a delta function in a bandwidth-limited plane-wave basis. Each function $D_i(\mathbf{r}) = D(\mathbf{r}-\mathbf{r}_i)$ is centred on a grid point $\mathbf{r}_i$, and the NGWFs are then written

$$\phi_\alpha(\mathbf{r}) \;=\; \sum_{|\mathbf{r}_i-\mathbf{R}_\alpha|<R_\phi} c_{i\alpha} D_i(\mathbf{r})$$

where $c_{i\alpha}$ are coefficients which are non-zero only for grid points $\mathbf{r}_i$ within a sphere of radius R_ϕ around the atom on which NGWF α is centred. The NGWFs are thus strictly localised within this sphere.

The spacing of the real-space grid is determined by an energy cutoff $E_{\rm cut}$ which determines the kinetic energy of the largest G-vectors, as in plane-wave pseudopotential approaches. In this work, $E_{\rm cut} = 400\,\mathrm{eV}$ was employed throughout. Four NGWFs per atom were used for Ga and As (initialised to $s-$and p-like atomic orbitals), and one NGWF per atom for H (initialised to s).

Large NGWF radii of $R_\phi = 10\,a_0$ were employed to ensure well-converged electronic structure and accurate forces. The ion cores were represented using norm-conserving pseudopotentials with three and five valence electrons respectively for Ga and As. The non-linear core correction approach was used to include core-valence interactions [25], which are important for both Ga and As. The local density approximation was used for exchange and correlation.

Due to the strict localisation of the NGWFs, the density $n(\mathbf{r}) = \sum_{\alpha\beta} \phi_\alpha(\mathbf{r}) K^{\alpha\beta} \phi_\beta(\mathbf{r})$, the overlap matrix $S_{\alpha\beta} = \langle\phi_\alpha|\phi_\beta\rangle$ and the Hamiltonian $H_{\alpha\beta} = \langle\phi_\alpha|\hat{H}|\phi_\beta\rangle$, can all be constructed in linear-scaling computational effort, through the use of the 'FFT box' approximation [26]. Minimisation of the total energy $E_T = \sum_{\alpha\beta} K^{\alpha\beta} H_{\beta\alpha} - E_{\mathrm{dc}}[n]$ then proceeds via a double loop: the outer loop optimises the spatial form of the NGWFs $\phi_\alpha(\mathbf{r})$ via a conjugate gradients algorithm, using the energy gradient $\partial E_T/\partial\phi(\mathbf{r})$ to construct the search direction. The inner loop optimises the matrix elements of the density kernel $K^{\alpha\beta}$. The NGWF optimisation has the effect of producing an *in-situ optimised*, minimally-sized, local orbital representation with an accuracy demonstrably equivalent to that of a plane-wave basis [27]. Sparse matrix algebra operations are performed using an atom-blocked scheme with sparsity determined by the overlaps of the local orbitals and by a real-space cutoff for the density kernel. Efficiently parallelised sparse matrix algebra routines [28, 29] mean that the calculations scale well to hundreds of parellel cores: most of the current results were performed using 256 Intel Core i7 cores (64 processors).

The kernel is optimised using a combination of methods [30] including LNV [31] and penalty functional minimisation [32], designed to ensure that the energy is minimised with respect to $K^{\alpha\beta}$ subject to the requirements that the density matrix is correctly normalised, in that $\mathrm{Tr}[\hat{\rho}] = N_e$, the number of electrons, and that it idempotent, in that $\hat{\rho}^2 = \hat{\rho}$. This has the effect of ensuring integer occupancies of zero or one for all single-electron eigenstates of the Hamiltonian, which is appropriate in an bulk insulator with a clearly defined bandgap or in an isolated system with a clear HOMO-LUMO gap. However, in the nanorod systems studied in this work there is often no significant HOMO-LUMO gap in the full density of states, which at first sight might question the appropriateness of a scheme employing integer occupation numbers. Fortunately in these cases there is always a clear gap in the local density of states everywhere in the rod, and the HOMO and LUMO have vanishing overlap due to being localised at opposite ends of the rod.

In these nanorods with widely separated polar surfaces, a density matrix corresponding to integer occupancies of the polar surface states is then more appropriate than one with 'smeared' metallic occupancies. In Ref. [24] we investigated the effect of smearing of occupancies around the Fermi level by performing test calculations on a few smaller rods with explicit diagonalisation of the Hamiltonian, followed by occupation of the states according to a Fermi-Dirac distribution with a self-consistently determined chemical potential. In all cases tried, while there was minor variation of the total energy, due to the slightly different occupations, there was not any major variation of the dipole moments observed (under 5% change in all cases).

2.2. Electrostatic Interactions

The strict localisation of the NGWFs means that in a large simulation cell, regions of vacuum at a significant distance from all the atoms (greater than R_ϕ) are not within the localisation region of any NGWFs. This means that apart from the 'whole-cell' operations such as calculation of the Hartree and XC energies, little extra computational effort is associated with the description of extra vacuum in the simulation cell. In a periodic simulation, it is therefore simple to pad a nanostructure with sufficient vacuum to prevent the wavefunctions of periodic images from actually overlapping.

However, the long-range of the Coulomb interaction, combined with the large dipole moments encountered in these systems, means that despite extra vacuum being free, it is still not feasible to converge electrostatic finite size effects with respect to simulation cell size in a periodic simulation.

Journal of Physics: Conference Series **367** (2012) 012002 doi:10.1088/1742-6596/367/1/012002

We therefore truncate the Coulomb interaction with a real-space cutoff [33], of the general form:

$$V(\mathbf{r},\mathbf{r}') = \frac{\Theta(R_c - |\mathbf{r}-\mathbf{r}'|)}{|\mathbf{r}-\mathbf{r}'|} ,$$

where $\Theta(r)$ is the Heaviside step function. In practice, it makes more sense in these elongated systems to employ a cutoff on the surface of a cylinder rather than a sphere. The method for achieving this has been detailed elsewhere [34].

The truncation of the Coulomb interaction automatically negates all multipole terms in the interaction between different nanorods. Of particular importance is that it prevents dipole-dipole term, which would otherwise act to damp the magnitude of the dipole that would be observed, as seen in Ref. [34].

2.3. Geometry Optimisation

The structures of the nanocrystals were determined using geometry optimisation with the BFGS algorithm. Positions were initialised to atomic positions as cleaved from the perfect crystal structure (itself determined by geometry optimisation of the comparable bulk wurtzite structure using the CASTEP plane-wave DFT package [35]). Initial positions of the surface terminating atoms were determined by choosing a typical bond-length perpendicular to the surface. The non-centrosymmetric nature of the Wurtzite structure means that surfaces perpendicular to the c-axis necessarily exposes a plane of Ga atoms on one polar surface and a plane of As atoms on the other. We will refer to these as the 'Ga surface' and 'As surface' respectively, regardless of the ligand species attached (eg H). The rods relaxed in this process are of length $\sim 3.5\,\text{nm}$ and width $\sim 1.2\,\text{nm}$.

We employ a shorthand notation to denote the possible types of terminating species. In this notation, 'H' is hydrogen, 'B' is a bare (unterminated) surface, and 'P' is 'pseudo-hydrogen', is a hydrogen-like nucleus with a variable nuclear charge to account for terminating ligands of varying electronegativity: $(5/4)e$ neutralises the Ga surfaces, and $(3/4)e$ neutralises the As surface [36]. The first letter indicates the lateral surface termination, and the letter after the slash indicates the polar surface termination.

For large nanorods this generated initial configurations with relatively low forces which relaxed in around 50-100 BFGS steps depending on the surface termination. Convergence was defined as when all forces were relaxed to under $0.1\,\text{eV/Å}$, positions were converged to within 0.005Å and the energy gain per step was under $0.03\,\text{meV/atom}$.

2.4. Local Density of States

One of the most useful means to understand and interpret the information provided by an electronic structure calculation is a description encompassing both the spatial and energetic distribution of the single-particle eigenstates. In the local-orbital framework provided by ONETEP this is achieved by performing a single diagonalisation of the Hamiltonian matrix in the NGWF basis, at the end of the calculation once self-consistency has been achieved. While this comes with a $O(N^3)$ computational cost, the prefactor is low because the NGWF basis is so small (four NGWFs per atom for Ga and As). Therefore, this diagonalisation remains feasible up to system sizes of around 10000 atoms, particularly if a parallel eigensolver such as ScaLAPACK is used. The generalised eigenproblem to be solved is of the form

$$\sum_\beta H_{\alpha\beta} M^{\beta}{}_{n} = \epsilon_n \sum_\beta S_{\alpha\beta} M^{\beta}{}_{n} \qquad (2)$$

The matrix $M^{\beta}{}_{n}$ describes the eigenvectors, which take the form $|\psi_n\rangle = \sum_\beta M^{\beta}{}_{n} |\phi_\beta\rangle$. Eq. 2 can be solved using the ScaLAPACK routine PDSYGVX, to obtain the eigenvalues $\{\epsilon_n\}$ and eigenvectors ${M^{\beta}}_{n}$. From these, we can obtain the density of states as

$$D(\epsilon) = \sum_{n} \delta(\epsilon - \epsilon_n) \, , \tag{3}$$

where in practice we replace the delta function with a Gaussian broadening of 0.1 eV. The local density of states in a given region I is calculated by including a projection of each eigenstate onto the local orbitals of region I, as

$$D_I(\epsilon) = \sum_{n} \delta(\epsilon - \epsilon_n) \, \langle\psi_n| \sum_{\alpha\in I} \left(|\phi^\alpha\rangle\langle\phi_\alpha|\right) |\psi_n\rangle. \tag{4}$$

The non-orthogonality of the projectors has been taken into account by raising the index of one of the terms, by implicitly constructing the contravariant dual of the NGWF (as in [37, 38]). The functions $|\phi^\alpha\rangle$ need not be explicitly constructed in real space, since the relationship $\langle\phi_\alpha|\phi^\beta\rangle = \delta_{\alpha\beta}$ implies that we can re-write Eq. 4 as follows:

$$\begin{aligned} D_I(\epsilon) &= \sum_{n} \delta(\epsilon - \epsilon_n) \sum_{\beta,\gamma,\alpha\in I} {(M^\dagger)_n}^{\gamma} \langle\phi_\gamma| \left(|\phi^\alpha\rangle\langle\phi_\alpha|\right) |\phi_\beta\rangle {M^\beta}_n \\ &= \sum_{n} \delta(\epsilon - \epsilon_n) \sum_{\alpha\in I} {(M^\dagger)_n}^{\alpha} \left(\sum_{\beta} S_{\alpha\beta} {M^\beta}_n\right) \end{aligned} \tag{5}$$

For these nanorod systems, a natural choice of regions over which to decompose the DOS is sets of slabs of atoms along the c-axis. Atoms are thus decomposed into groups according to their position along c, and a separate LDOS curve obtained for each group (as in Figure 2).

2.5. *Calculating Dipole Moments*

The total polarisation of an isolated nanostructure can be expressed as the sum of the ionic and electronic polarisation. The ionic polarisation is simply

$$\mathbf{d}_{\text{ion}} = \sum_{I} Z_I \mathbf{R}_I \, . \tag{6}$$

The electronic polarisation $\mathbf{d}_{\text{el}}$, meanwhile, is given in an isolated system by

$$\mathbf{d}_{\text{el}} = -\int \mathbf{r} n(\mathbf{r}) \, d\mathbf{r} \, . \tag{7}$$

To evaluate this in the local orbital framework, we first evaluate the matrix elements of the position operator in the NGWF representation, $\mathbf{R}_{\alpha\beta} = \langle\phi_\alpha|\mathbf{r}|\phi_\beta\rangle$, within the FFT box (adjusting for the origin of the FFT box so as to give a result relative to the origin of the simulation cell). Then we write Eq. 7 as

$$\mathbf{d}_{\text{el}} = -\int \mathbf{r}\phi_\alpha(\mathbf{r}) K^{\alpha\beta} \phi_\beta(\mathbf{r}) \, d\mathbf{r} = K^{\alpha\beta} \mathbf{R}_{\beta\alpha} \, ,$$

indicating that the polarisation can be calculated directly from the density kernel and the NGWFs, without requiring any knowledge of the individual eigenstates. This allows dipole moments to be calculated easily for extremely large systems.

The end result is the total polarisation

$$\mathbf{d}_{\text{tot}} = \mathbf{d}_{\text{el}} + \mathbf{d}_{\text{ion}} \, , \tag{8}$$

which for a charge-neutral system is independent of the origin used for the position operator as long as it is the same in both terms.

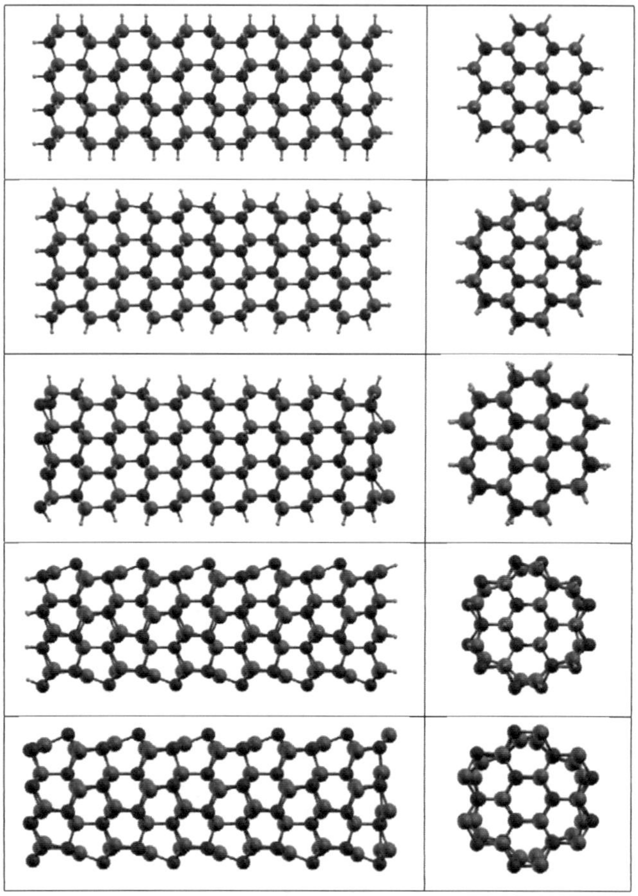

Figure 1. Nanorods before and after geometry optimisation. Arsenic atoms are blue and Gallium atoms are red. From top: Unrelaxed rod; H/H rod; H/B rod; B/H rod; B/B rod.

Below: Table of dipole moments d_z (in Debye) before and after geometry relaxation, for the rod types shown (reproduced from Ref. [23]). Dipole moments change quantitatively, but qualitative trends are unchanged in all cases except B/H.

Rod	d_z Before	d_z After
H/H	-131.6	-103.0
H/B	+55.2	+68.0
B/H	-22.8	-89.2
B/B	+15.4	+25.5

3. Results

3.1. Effect of Surface Relaxation on Nanorod Dipole Moment

Figure 1 shows the atomic configurations before and after relaxation for four different choices of surface termination. It can be seen that in the case of H-termination on the polar and lateral surfaces, the unrelaxed structure is not significantly altered, except that the As-H and Ga-H bonds align with where the next atom in the bulk structure would be, to restore the tetrahedral configuration of each Ga and As site, as would be expected. The differing electronegativities of Ga and As also mean that the Ga-As bonds in the surface layer tilt somewhat. In the cases with bare surfaces, there is much more significant relaxation. When the lateral surfaces are unterminated, a major distortion of the surface layer occurs so as to produce flat, threefold coordinated layers of Ga and As atoms, with one ridge occurring per primitive cell in the $c-$direction. This has the effect of partially 'healing' the local bandgap, restoring overall insulating behaviour.

A bare polar surface is seen to be similarly energetically unfavourable: surface atoms of both species are again flattened into planes, so as to result in charge-neutral surface. Excess atoms appear to have a tendency towards being expelled from the structure, although the relaxation scheme employed here does not actually cause them to do so in practice. Clearly, in some cases, this suggests that the structures considered here will not be stable. A semi-infinite polar surface with similar termination would be unstable to reconstructions involving changes in the composition of the surface (addition or removal of atoms). At some size, therefore, one would expect the same thing to happen even for a finite-sized nanorod. This possibility is not accounted for within the current model, but remains an important issue for future work to investigate.

Journal of Physics: Conference Series **367** (2012) 012002 doi:10.1088/1742-6596/367/1/012002

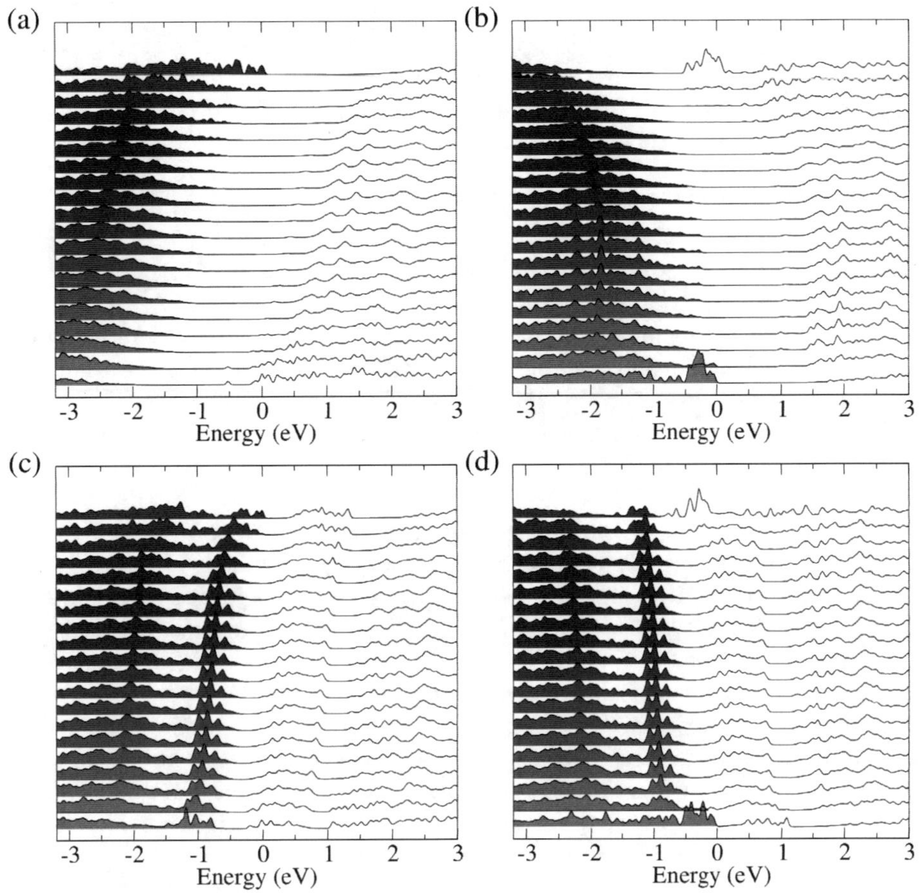

Figure 2. Extract of local densities of states around the valence band edge decomposed into slabs along the c-direction (each slab offset vertically in figures), for various rod terminations: a) H/H, b) H/B, c) B/H and d) B/B. Filled curves show the local density of states below the gap for each slab; the line shows the local density of empty states above the gap for each slab. The fill color varies from blue on the As(-H) terminated end to red on the Ga(-H) terminated end.

Figure 1, in particular the dipole moments of the table, suggest that to obtain a qualitative understanding of the dependence of the dipole moment on the surface termination it is not necessary to relax the surface. In fact, the variations in electronic structure between rods with different surface species can best be understood by considering very large unrelaxed nanorods with pristine unreconstructed surfaces.

3.2. Variation of Dipole Moments with Surface Termination
Figure 2, presents the local density of states of four simulated nanorods of length $\sim 12\,\mathrm{nm}$ and width $\sim 2\,\mathrm{nm}$, varying in the surface species with which dangling bonds are terminated. It shows that, as might be expected, the presence of a strong electrostatic field shifts the peaks of the density of states up in energy at the negatively-charged end and down at the positively charged end. The most notable feature, discussed but not fully explained in Ref. [23], is that the dipole moment itself clearly does not depend strongly on the underlying polarisation of the wurtzite structure, but depends very strongly on the surface termination of both the polar and lateral surfaces. This dependence can be sufficient to reverse the direction of polarisation (hence the

Journal of Physics: Conference Series **367** (2012) 012002 doi:10.1088/1742-6596/367/1/012002

Nanorod	H/H	H/B	B/H	B/B
Dipole moment d_z	−614	+330	−235	+125

Table 1. Dipole moments (in Debye) of the four types of rod whose LDOS are shown in Fig. 2.

oppositely-directed electric fields in Figs 2(b) and 2(d) compared to 2(a) and 2(c)).

Table 1 shows the variation of the dipole moment between the four types of rod. This variation cannot be explained simply in terms of the differing ionic charges of the terminating species: arguments based on counting ionic charges and the charge associated with Ga-As, Ga-H and As-H bonds can be used to estimate roughly what change in polarisation might be expected to result from a given change in surface termination, in the absence of electronic or ionic polarisation (i.e. what would be obtained simply by cleaving the bulk structure without allowing electron density to relax). These uniformly predict both that changes in the dipole moments of the rods would be considerably larger than those observed, and do not explain even qualitatively the variations seen in Fig. 2.

Examining the different figures, the most obvious effect the variation of surface termination has on the electronic structure is through the position and number of surface states in the bandgap. It can be observed that in all cases, the polarisation and the corresponding voltage drop along the rod are such that states adjacent to the valence band edge at one polar surface are aligned in energy with states adjacent to the conduction band at the other polar surface. However, this occurs in a different way depending on the nature of the surface termination: in the rods with H-terminated polar surfaces, the surface-induced states specific to the polar surfaces (i.e. in the top and bottom curves) consist of redistribution of weight in the DOS around the valence band maximum on the Ga surface and around the conduction band maximum on the As surface. Alignment between these states can thus be achieved by an electric field which raises the potential at the Ga surface and and lowers it at the As surface.

Conversely, for the rods with bare (unterminated) polar surfaces, the surface-induced 'dangling-bond' states specific to the polar surface are adjacent to the conduction band on the Ga surface and the valence band at the As surface, and are much more strongly peaked around a specific energy. This means an electric field of opposite sign is necessary to bring the surface states into energetic alignment.

When one examines the rods without lateral-surface termination, Figs. 2(c) and 2(d), one sees a significant density of states associated with the dangling bonds on the lateral surfaces adjacent to the valence band maximum: these have the effect of almost closing the bandgap.

In summary, therefore, one can explain the trends of Table 1 in terms of an argument based on pinning of the Fermi level as follows: exposing layers of Ga on one face and As on the other necessarily creates a charge imbalance between opposite polar surfaces, and thus a dipole forms. However, if this dipole moment were too large, unoccupied electronic states would be pushed below the Fermi level on the positively charged end and raised above it on the negatively charged end. This would have two effects: a higher-energy configuration overall, and a reversal of the dipole moment due to the charge transfer. However, the bulk of the rod is itself polarisable, so instead charge flows to oppose the field by polarising the bonds along the rod axis until the positions of the filled and empty states are aligned at opposite ends of the rod. The dipole moment is thus limited in magnitude by Fermi level pinning, which effectively ensures that the in the final electronic configuration, the voltage drop between the ends of the rod cannot exceed the difference between the energies of the occupied states at one end and the unoccupied states at the other end.

The differences in sign and magnitude of the dipole moments can be explained in terms of the effects of the surface species on the local density of states at the surfaces: bare lateral surfaces significantly close the bandgap, resulting in a lower dipole moment for H/B and B/B compared to H/H and B/H respectively. Meanwhile, going from H-termination to bare surfaces at the polar ends switches the location of the occupied states, moving them from the Ga end to the As end, and vice-versa for the empty states. Therefore the sign of the dipole moment switches from negative to positive, while the smaller gap resulting from the position of the dangling-bond states inside the local gap at each end ensures the magnitude reduces.

4. Conclusions

We have presented in this paper a methodology for electronic structure calculations of nanostructures, where the computational effort required scales linearly with system size, which enables determination of the structural and electronic properties calculations of whole semiconductor nanocrystals of realistic size (up to 2862 atoms). In our earlier work on this subject [23], we described and examined the factors that determine the charge distribution in GaAs nanorods with varying surface termination. In subsequent work [24] we introduced the importance of the concept of Fermi Level Pinning in describing the charge distribution in nanorod systems, and explained how it could be used to explain the variation of dipole moment with size, shape, and composition of the nanorod. In the current work we have applied the concept of FLP to variation of dipole moment with surface termination, and showed how it can be used to explain the results of the former paper.

This work was supported by EPSRC (UK) under Grant No. EP/G05567X/1 (NDMH), EPSRC (PWA), the EC under Contract No. MIRG-CT-2007-208858 (PT), and a Royal Society University Research Fellowship (PDH). All calculations were run on the Imperial College HPC Service.

References

[1] N. Tessler, V. Medvedev, M. Kazes, S. Kan, and U. Banin, Science **295**, 1506 (2002).
[2] M. Kazes, D. Y. Lewis, Y. Ebenstein, T. Mokari, and U. Banin, Adv. Mater. **14**, 317 (2002).
[3] W. U. Huynh, J. J. Dittmer and A. P. Alivisatos, Science **295**, 2425 (2002).
[4] M. A. El-Sayed, Accounts Chem. Res. **37**, 326 (2004).
[5] X. Michalet, F. F. Pinaud, L. A. Bentolila, J. M. Tsay, S. Doose, J. J. Li, G. Sundaresan, A. M. Wu, S. S. Gambhir, and S. Weiss, Science **307**, 538 (2005).
[6] E. V. Shevchenko, D. V. Talapin, N. A. Kotov, S. O'Brien, and C. B. Murray, Nature **439**, 55 (2006).
[7] Z. Nie, A. Petukhova, and E. Kumacheva, Nat. Nanotechnol. **5**, 15 (2010).
[8] M. Heiss, *et. al.*, Phys. Rev. B **83**, 045303 (2011).
[9] S. A. Blanton, R. L. Leheny, M. A. Hines, and P. Guyot-Sionnest, Phys. Rev. Lett., **79**, 865 (1997).
[10] M. Shim and P. Guyot-Sionnest, J. Chem. Phys. **111**, 6955 (1999).
[11] L.-S. Li and A. P. Alivisatos, Phys. Rev. Lett. **90**, 097402 (2003).
[12] G. Galli, M. Parrinello, Phys. Rev. Lett. **69**, 3547 (1992).
[13] S. Goedecker, Rev. Mod. Phys. **71**, 1085 (1999).
[14] J. M. Soler, E. Artacho, J. D. Gale, A. García, J. Junquera, P. Ordejón and D. Sánchez-Portal, J. Phys.: Condens. Matter **14**, 2745-2779 (2002).
[15] D. R. Bowler, R. Choudhury, M. J. Gillan, T. Miyazaki, phys. stat. sol. (b) **243**, 989 (2006).
[16] J. VandeVondele, M. Krack, F. Mohammed, M. Parrinello, T. Chassaing and J. Hutter, Comp. Phys. Commun. **167**, 103 (2005).
[17] J.-L. Fattebert and J. Bernholc, Phys. Rev. B **62**, 1713 (2000).
[18] J.-L. Fattebert and F. Gygi, Phys. Rev. B **73**, 115124 (2006).
[19] M. Challacombe, J. Chem. Phys. **110**, 2332 (1999).
[20] T. Ozaki and H. Kino, Phys. Rev. B **72**, 045121 (2005); http://www.openmx-square.org/.
[21] C.-K. Skylaris, P. D. Haynes, A. A. Mostofi, and M. C. Payne, J. Chem. Phys. **122**, 084119 (2005).
[22] N. D. M. Hine, M. Robinson, P. D. Haynes, C.-K. Skylaris, M. C. Payne, and A. A. Mostofi, Phys. Rev. B **83**, 195102 (2011).
[23] P. W. Avraam, N. D. M. Hine, P. Tangney, and P. D. Haynes, Phys. Rev. B **83**, 241402(R) (2011).
[24] P. W. Avraam, N. D. M. Hine, P. Tangney, and P. D. Haynes, Phys. Rev. B **85**, 115404 (2012).
[25] S. G. Louie, S. Froyen, and M. L. Cohen, Phys. Rev. B **26**, 1738 (1982).

[26] C.-K. Skylaris, A. A. Mostofi, P. D. Haynes, O. Diéguez and M. C. Payne, Phys. Rev. B **66**, 035119 (2002).
[27] C.-K. Skylaris and P. D. Haynes, J. Chem. Phys. **127**, 164712 (2007).
[28] N. D. M. Hine, P. D. Haynes, A. A. Mostofi, and M.C. Payne, Comp. Phys. Commun. **180**, 1041 (2009).
[29] N. D. M. Hine, P. D. Haynes, A. A. Mostofi, C.-K. Skylaris and M. C. Payne, J. Chem. Phys. **134**, 114111 (2010).
[30] P.D. Haynes, C.-K. Skylaris, A.A. Mostofi and M.C. Payne, J. Phys.: Cond. Mat. **20**, 294207 (2008).
[31] X.-P. Li, R. W. Nunes and D. Vanderbilt, Phys. Rev. B **47**, 10891 (1993).
[32] P. D. Haynes and M. C. Payne, Phys. Rev. B **59**, 12173 (1999).
[33] C. A. Rozzi, D. Varsano, A. Marini, E. K. U. Gross, and A. Rubio, Phys. Rev. B **73**, 205119 (2006).
[34] N. D. M. Hine, J. Dziedzic, P. D. Haynes, and C.-K. Skylaris, J. Chem. Phys. **135**, 204103 (2011).
[35] S. J. Clark, M. D. Segall, C. J. Pickard, P. J. Hasnip, M. I. J. Probert, K. Refson, and M. C. Payne, Z. Kristallogr. **220** 567570 (2005).
[36] X. Huang, E. Lindgren, and J. R. Chelikowsky, Phys. Rev. B **71**, 165328 (2005).
[37] D. D. O'Regan, M. C. Payne and A. A. Mostofi, Phys. Rev. B **83**, 245124 (2011).
[38] D. D. O'Regan, N. D. M. Hine, M. C. Payne and A. A. Mostofi, Phys. Rev. B **85**, 085107 (2012).

A study of atomic orbital basis sets for doped silicon nanowires

***D Sharma[1] and G Fagas**
Tyndall National Institute, University College Cork, Cork, Ireland
*also at the Department of Chemistry, University College Cork, Cork, Ireland
E-mail: dimpy.sharma@tyndall.ie, georgios.fagas@tyndall.ie

Abstract. First-principle calculations are performed within Density Functional Theory to determine the accuracy of numerical atomic orbital basis sets used for studies of highly-doped silicon nanowires. Boron, phosphorous, gallium and arsenic atoms are used as dopants. Structural and electronic properties are studied and the performance of optimised basis sets is benchmarked. A good compromise between efficiency and accuracy is offered by optimised double zeta polarised basis sets. The transferability of numerical atomic orbitals for studies of silicon nanowires is tested by applying in nanowires with different orientations.

1. Introduction

One-dimensional semiconductor nanostructures, particularly silicon nanowires (SiNWs), have attracted extensive research efforts over the past decade [1]. In recent device applications SiNWs cover a wide range of areas such as transistors for light emitting diodes (LEDs), ballistic field-effect transistors (FETs) and sensitive nanosensors. SiNW solar cells are expected to increase absorption of light, thereby, yielding higher efficiencies at lower cost. One can change the electronic and optical properties properties through doping of the SiNWs [2, 3] as demonstrated by evidence on tuneable control of their conductivity and photoluminescence [4, 5]. Therefore a detailed understanding of the structural and electronic properties of doped SiNWs is important. In particular, there is a growing need to address large-scale device simulations from first principles and a good compromise between efficiency and accuracy is sought [6]. To this end, atomic orbitals offer tractable basis sets but a systematic approach to improve their predicting capabilities is needed prior to applying them in practice [7, 8].

Here, our aim is to discuss quantitative estimates in SiNWs derived from Density Functional Theory (DFT) implementations using numerical atomic orbitals (NAOs) [9]. We employ Density Functional Theory to investigate the electronic structure of [110]-oriented hydrogenated SiNWs with diameters 1.15nm and highly-doped with boron, phosphorous, arsenic and gallium. These nanostructures are particularly relevant for the recent proposal of junctionless transistors [10] where doping concentrations as high as 10^{21} are required. We report on the convergence of different contractions of primitive and optimised NAOs by studying the structural and electronic properties of SiNWs with an uneven distribution of dopants.

---

1 dimpy.sharma@tyndall.ie

(a) 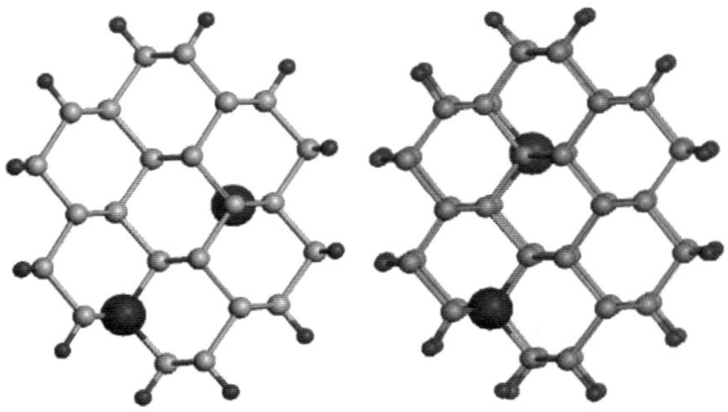 (b)

Figure 1: Cross section of [110]-oriented geometrically optimised doped silicon nanowires with dopants distributed at two different uneven locations assigned as (a) configuration 1 and (b) configuration 2.

2. Methodology

We perform DFT calculations within a generalized gradient approximation (GGA) to the exchange and correlation functional. Norm-conserving fully relativistic pseudopotentials and numerical atomic basis functions are generated by the software package ADPACK [13]. The systematic convergence of the quality of the basis sets is performed by increasing the number of primitive orbitals describing the valence electrons and by adding higher angular momentum functions (d-polarisation). A Monkhorst-Pack k-point sampling on a 15x1x1 grid in the Brillouin zone along the nanowire axis direction is used. The SiNW is put in a supercell with more than 24 Å vacuum spacings in the lateral directions to avoid any interactions between the neighbouring image nanowires. The SiNWs and the unit cell lattices are fully relaxed. Structural optimisation calculations are performed until the absolute value of force acting on each atom is 0.01eV/Å.

Substitutional doping of hydrogen passivated SiNWs is achieved by exchanging two silicon atoms in the supercell with B, P, Ga and As atoms. This yields a doping concentration of 4.34 % .We have taken two different configurations of each doped SiNW by placing dopants at uneven sites and have assigned them as configuration 1 and configuration 2. Figure 1 shows the example of B-doped geometry optimised nanowires. We compare the total energy and electronic structure calculated using a hierarchy of basis sets, from single-ζ minimal to multiple-ζ with polarization orbitals. The notation snmpn'm'dn" of Ozaki et al [11, 12], and as implemented in the OpenMX DFT code [13], is used for indicating the abbreviation of numerical atomic orbital basis. For example, s32p32d1 means that two optimised orbitals are constructed from three primitive functions to describe s and p electrons respectively, while one contraction is employed for the d orbitals. The basis size for the silicon and dopant atoms was varied while keeping a minimal basis set (s1) for H. The minimal basis set for Ga and As is s1p1d1 and the starting point is to include d-orbital polarisation functions also for Si.

3. Results and Discussions

The calculated total energy for the various NAO basis sets reduces with increasing the basis size and/or applying orbital optimisation as can been seen in Figure 2. By increasing the contraction and adding d-orbital polarisation functions, the total energy per atom converges to less than 0.01eV per atom. This is a minimum requirement regarding the basis set sufficiency of NAOs and it is demonstrated here for the extended system of SiNWs. The respective convergent values for the NAO basis sets are 2.04eV, 1.89eV, 1.63eV and 2.46eV for SiNWs doped with B, P, As and Ga, respectively. The optimisation of orbitals overall yields significant improvement and it becomes less important as the number of contractions increases as may be expected unless explicitly indicated, there is no difference between both the configurations of each system.

A simple inspection of Figure 2 may suggest that optimised orbitals do not yield significant improvements when multiple- ζ basis sets are used. This may hold well for the structural properties. But a previous comparison with plane wave benchmark calculations in hydrogenated and hydroxylated

Journal of Physics: Conference Series **367** (2012) 012003 doi:10.1088/1742-6596/367/1/012003

SiNWs [14] has indicated that electronic structure features and transport properties derived thereafter greatly improve by orbital optimisation. This is demonstrated in Figure 2 by comparing the band structures between primitive and optimised double-ζ polarised NAO for the first few valence and conduction sub-bands. The sensitivity to orbital optimisation is lesser for the valence electronic states.

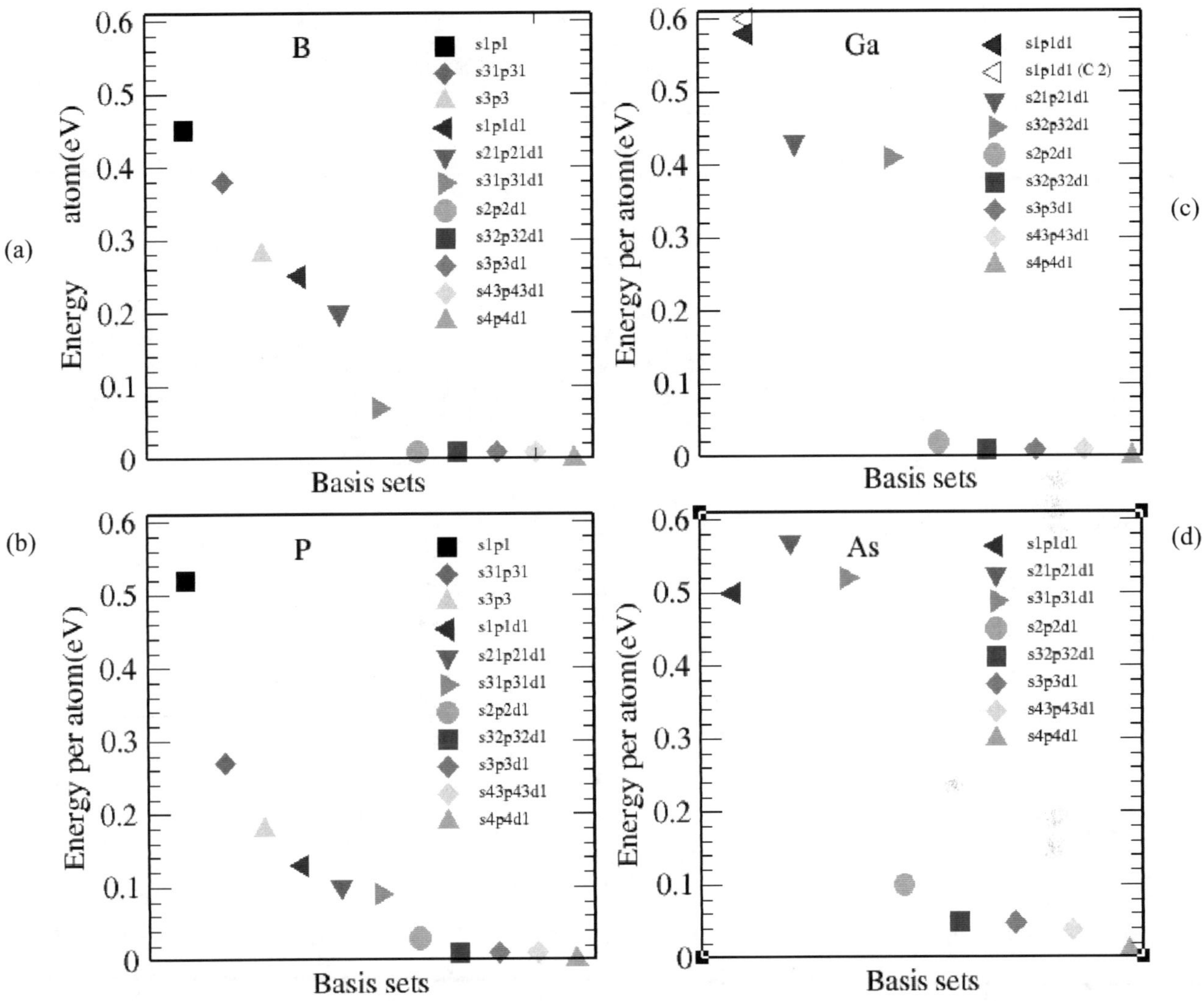

Figure 2: Variation of the total energy per atom in [110]-oriented silicon nanowires for different numerical atomic basis and dopant atoms, that is, (a) boron doped ($Si_{46}B_2H_{32}$), (b) phosphorous doped ($Si_{46}P_2H_{32}$), (c) gallium doped ($Si_{46}Ga_2H_{32}$) and (d) arsenic doped ($Si_{46}As_2H_{32}$). All values are referenced to the calculation using the s4p4d1 NAOs.

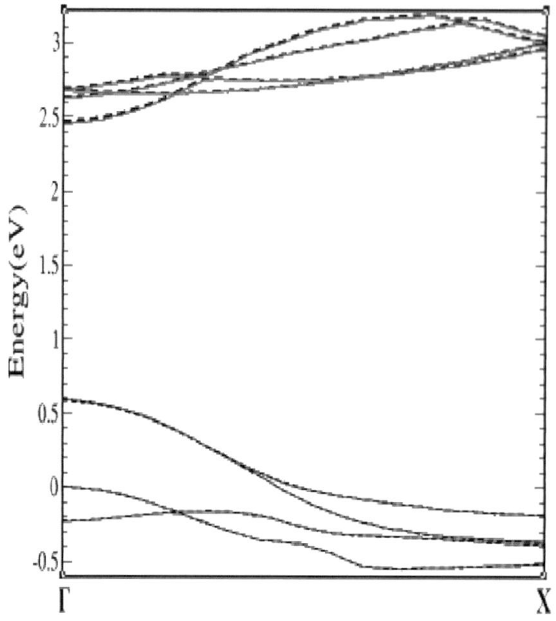

(a)

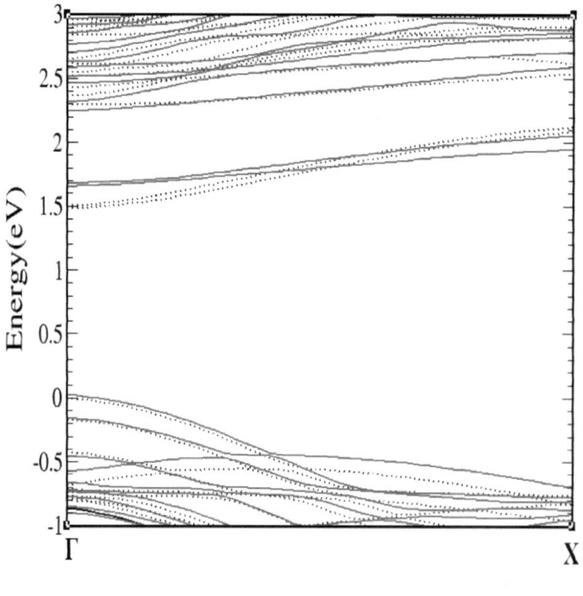

(b)

Figure 3: Comparison of the first few valence and conduction (sub-)bands of (a) gallium doped ($Si_{46}Ga_2H_{32}$) and (b) arsenic doped silicon nanowire ($Si_{46}As_2H_{32}$) calculated using primitive (red solid lines) and optimized double-zeta polarized NAOs (black solid lines).

Finally, the transferability of optimised orbitals is explored by two types of control calculations. First, we consider phosphorus doped SiNWs with three different orientations [100], [110] and [111]. We then exchange the NAOs as obtained from the various systems and compare their electronic structure. For example, we took the contracted orbitals obtained after orbital optimisation in [110]-oriented P-doped SiNW as inputs for calculations with [100]- and [111]-oriented P-doped SiNWs. For a fixed orientation, the total energies obtained after geometry relaxation are equal for the various NAO basis sets independent on the wire orientation used for the orbital optimisation. This procedure also yields almost identical band structures. The result of our comparison for the [100] orientation is shown in Figure 4. This procedure can prove very efficient in larger nanowires (or indeed other systems) where orbitals can be optimised in a small controlled environment and then transferred to the bigger system.

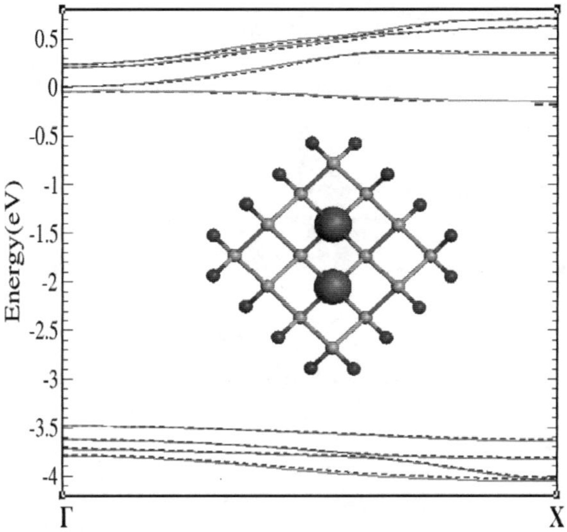

Figure 4: Transferability comparison of optimised numerical atomic orbitals. The band structure of a [100]-oriented P-doped is shown as calculated using orbitals obtained either from optimisation within the same structure (magenta solid lines) or from optimisation performed on [110]-oriented nanowires (blue dashed lines).

Concluding remarks

In summary, we evaluated the implementation of numerical atomic basis sets by studying the total energy and electronic bands of silicon nanowires with hydrogen surface terminations and doped with B, P, As and Ga atoms. It has been shown that there is systematic improvement with respect to the contraction scheme. Whilst total energies may converge with increasing basis size, optimised NAOs offer a better option to study the electronic properties of SiNWs with fixed basis size. In particular, the optimised double-zeta polarized basis set offers a reasonable approximation. It was also shown that there is some transferability of optimised NAOs which can be used to address nanowire calculations in more complex (larger) environments. These results complement studies that put on a firm basis the use of NAOs as an efficient method, alternative to plane waves, to calculate the properties of silicon nanowires.

Acknowledgment

We would like to thank the Irish Centre for High-End Computing (ICHEC) for the availability of computational resources as. This work was supported by the Science Foundation Ireland grant number 06/IN.1/I857. We would also like to thank Hadi Arefi and Marios Iakovidis from the Electronics Theory group at Tyndall for their help throughout the work.

Journal of Physics: Conference Series **367** (2012) 012003 doi:10.1088/1742-6596/367/1/012003

References

[1] Cui Y, Lieber C M 2001 Science 291, pp. 851-853
[2] Wang Y, Lew K K, Ho T T, Pan L, Novak S , Dickey E C, Redwing J M, Mayer T S 2005 *Nano Lett.* **5**, p 2139.
[3] Vallett A L, Minassian S, Kaszuba P, Datta S, Redwing J M, and Mayer T S 2010, *Nano Lett.* **10**, pp 4813–4818
[4] Fernandez-Serra M V, Adessi C, and Blase X 2006 *PRL* **96** pp. 166805
[5] Kayes B M, Atwater H A, Lewis N S 2005 J. Appl. Phys. **97** p. 114302
[6] Junquera J, Pazl O, Sánchez-Portal D and Artacho E 2001 Phys. Rev. B **64** p. 235111
[7] Gusso M, *J .Chem. Phys.*2008 **128**, p. 044102
[8] Driscoll J A and Varga K 2010 Phys. Rev. B **81**, 115412
[9] Kenny S D, Horsfield A P and Fujitani H 2000 Phys. Rev. B **62**, pp. 4899–4905
[10] Lida Ansari, Baruch Feldman, Giorgos Fagas, Jean-Pierre Colinge, and James C. Greer Appl. Phys. Lett. 2010 **97**, 062105
[11] T. Ozaki, Phys. Rev. B, 67 (2003) , p. 155108
[12] T. Ozaki, H. Kino, Phys. Rev.B, 69 (2004), p. 195113
[13]http://www.openmx-square.org/
[14] Sharma D, Arefi H H, and Fagas G *(review under process)*

Journal of Physics: Conference Series **367** (2012) 012004 doi:10.1088/1742-6596/367/1/012004

Density functional studies of the defect-induced electronic structure modifications in bilayer boronitrene

A M Ukpong[1,*] and N Chetty[1,2]
[1] Department of Physics, University of Pretoria, Hatfield 0002, South Africa
[2] National Institute for Theoretical Physics, Johannesburg 2000, South Africa

E-mail: aniekan.ukpong@up.ac.za

Abstract. The van der Waals interaction-corrected density functional theory is used in this study to investigate the formation, energetic stability, and inter-layer cohesion in bilayer hexagonal boronitrene. The effect of inter-layer separation on the electronic structure is systematically investigated. The formation and energetic stability of intrinsic defects are also investigated at the equilibrium inter-layer separation. It is found that nonstoichiometric defects, and their complexes, that induce excess nitrogen or excess boron, in each case, are relatively more stable in the atmosphere that corresponds to the excess atomic species. The modifications of the electronic structure due to formation of complexes are also investigated. It is shown that van der Waals density functional theory gives an improved description of the cohesive properties but not the electronic structure in bilayer boronitrene compared to other functionals. We identify energetically favourable topological defects that retain the energy gap in the electronic structure, and discuss their implications for band gap engineering in low-*n* layer boronitrene insulators. The relative strengths and weaknesses of the functionals in predicting the properties of bilayer boronitrene are also discussed.

1. Introduction

The hexagonal boron nitride (*h*-BN) monolayer has attracted sustained interests in recent times as the boron nitride analogue of graphene. Its stability as a freestanding monolayer and as a coupled system of few single layers (FSLs) [1] offers a unique opportunity for the exploration of the rich physics associated with graphene. Because *h*-BN has a wide band gap and no lattice mismatch with graphene, thin *h*-BN layers are useful in microelectronics as thin top-dielectric to gate graphene [2], inert substrates (or spacers) and scaffolds for graphene support [3,4]. For instance, the use of FSLs of *h*-BN in quantum-well heterostructures has been reported to improve the electronic quality of heterojunction devices compared to the use of highly-oriented pyrolytic boron nitride [5]. Nevertheless, the fewer number of experimental studies on *h*-BN nanosheets indicates that low-*n* layer *h*-BN is not fully explored compared to graphene. This situation arises because the synthesis of BN-based nanostructures is significantly more challenging than carbon based nanostructures because the well-developed methods for graphene synthesis are not always suitable for boronitrene synthesis [6]. Defects may be introduced deliberately to modify the wide band gap during synthesis. Because such

* To whom any correspondence should be addressed.

Journal of Physics: Conference Series **367** (2012) 012004 doi:10.1088/1742-6596/367/1/012004

defects can also occur spontaneously, it is important to understand the influence of defects on the electronic structure of low-*n* layers of *h*-BN. The dependence of the formation energy on growth conditions is also relevant for the stability of defective *h*-BN layers. This paper focuses on the modifications of the electronic structure in bilayer *h*-BN due to intrinsic defects. Defects that are capable of retaining the gap between the valence and conduction bands are identified.

2. Computational Methods

First principles calculations were performed based on density functional theory (DFT) as implemented in Quantum ESPRESSO [7]. The calculations were initially performed using the local density and generalized gradient approximations (LDA [8] and GGA [9]) to the exchange-correlation (XC) interactions. The GGA parameterization of Perdew, Burke, and Ernzerhof (GGA-PBE) [9] was used. Because of dispersive interactions [10] between coupled *h*-BN layers, we have also used the van der Waals interaction-corrected density functional (vdW-DF) to determine the ground state properties [11,12]. The interaction between valence electrons and ionic cores were described using the projector-augmented wave (PAW) potentials [13,14]. The kinetic energy cut-off of 500 eV was used for the plane wave expansion, and the electronic energy was converged to within 10^{-7} eV. The Monkhorst-Pack scheme was used to sample the Brillouin zone [15]. The electronic states were populated in accordance with the Fermi distribution function, with a smearing parameter of 0.2 eV. The atomic forces were calculated using the Hellman-Feynman theorem, and the atomic positions were relaxed until the forces were reduced to less than 0.03 eVÅ^{-1}. Convergence tests on the total energy differences were performed on the 4-atom unit cell with respect to k-points. It was concluded that the 12×12×1 sampling of the Brillouin zone was sufficiently converged. The total energy of the pristine bilayer was optimized with respect to the lattice parameter for the three XC functionals. The vacuum height was optimised to 15 Å to avoid spurious interlayer interactions.

3. Results and Discussion

Bilayer *h*-BN exists in five stacking sequences because the basis atoms of the top and bottom layer are dissimilar. Two of the sequences (*i.e.* AA and AA') give the simple stacking conformation, while the remainder (i.e. by A'B, AB' and AB) give the Bernal AB stacking conformation [16]. Table 1 shows the effect of stacking sequence on vdW-DF optimised properties of bilayer *h*-BN. The difference in total energy per unit cell δE is evaluated with respect to the AB-stacked structure because its total energy is lowest. From Table 1, the total energy of the AA'-stacked *h*-BN bilayer is 3.4 meV and 10.75 meV lower than the AB' and A'B-stacked bilayer *h*-BN structures respectively. We therefore conclude that the most unstable form of pristine *h*-BN bilayer is stacked in the AA sequence.

Table 1. Effect of stacking sequence on properties of bilayer h-BN. The last column indicates the pair of atoms used to determine the interlayer distance *d*.

Stacking	δE (meV/cell)	Δ_c (eV)	Δ_v (eV)	m^*	$l_{B\text{-}N}$ (Å)	d (Å)	Atom
A'B	36.87	0.77	1.353	0.36	1.451	3.79	N-N
AB'	29.52	0.96	1.210	0.29	1.443	3.36	B-B
AB	0.00	1.46	1.165	0.19	1.452	3.32	B-N
AA'	26.12	0.87	1.305	0.32	1.452	3.55	B-N
AA	37.28	0.58	1.208	0.48	1.452	3.74	B-B

The vdW-DF, GGA-PBE and LDA band structure of the AB-stacked pristine *h*-BN bilayer is shown in Figure 1. The functionals yield similar electronic structure and show non-degenerate π-bands at point K. However, the GGA and vdW-DF give small energy shifts (see Table 2), whereas the LDA gives a larger shift between the two top-most lying π-bands. The difference in the positions of the GGA-PBE and vdW-DF Fermi levels (indicated with dashed horizontal lines in Fig. 1) is small. This indicates that vdW-DF does not necessarily give an improved description of the electronic structure in bilayer *h*-BN. Although the vdW-DF band gap (4.76 eV) is significantly higher than the LDA band gap, the vdW-DF band gap is 0.74 eV lower than the experimental value of 5.50 eV [17]. By contrast, the deviation of the GGA-PBE band gap (5.35 eV) from the experimental band gap is less than 3%. We therefore conclude that vdW-DF does not provide an improved description of the electronic structure in the bilayer *h*-BN. Instead, GGA-PBE gives a better estimate of the experimental band gap.

Table 1 also shows the vdW-DF conduction and valence band widths. These were evaluated as differences, $\Delta = E(\text{K-point}) - E(\Gamma\text{-point})$, relative to the Fermi energy for the lowest unoccupied (Δ_c) and highest occupied (Δ_v) π-bands, respectively. From Δ_c (see Table 1), the predicted effective mass m^* is lowest (0.19) in the AB-stacked sequence and highest (0.48) in the AA-stacked sequence. The group velocity v_g for electron drift in the conduction band along the Γ→K direction of the Brillouin zone also shows that electrons attain their highest speed (54×10^6 m/s) and lowest speed (21×10^6 m/s) in the AA- and AB-stacked structures respectively. In the valence band however, electrons in the AB-stacked structure have the highest speed (27×10^6 m/s) compared to the lowest speed (23×10^6 m/s) in the A'B-stacked structure.

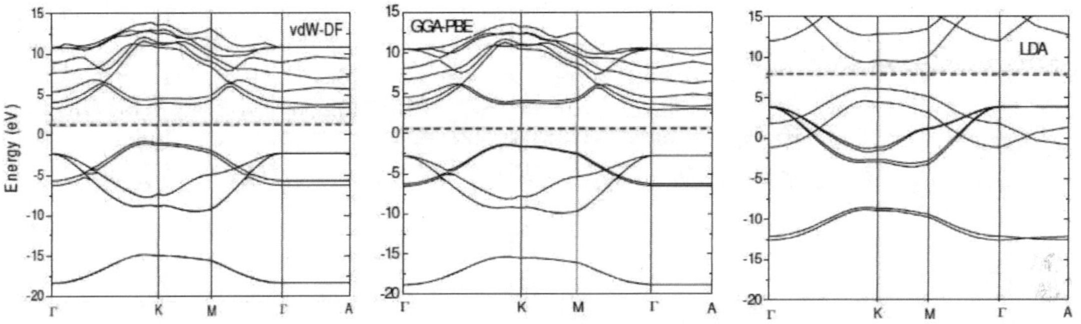

Figure 1. Band structure of pristine AB-stacked bilayer boronitrene at equilibrium interlayer spacing.

Table 2. Stereo-electronic properties of pristine bilayer hexagonal boronitrene.

Functional	$l_{\text{B-N}}$ (Å)	d_0(Å)	E_{coh} (eV/atom)	Electronic properties at point-K	
				Band gap (eV)	π-band shift (eV)
vdW-DF	1.442	3.32	-0.40	4.76	0.268
GGA-PBE	1.452	4.28	-0.33	5.35	0.094
LDA	1.447	3.78	-0.26	3.41	1.587

The lattice parameter is 2.50 Å, 2.52 Å, and 2.51 Å after optimisation with vdW-DF, GGA-PBE, and LDA functional, respectively. These values compare favourably with the experimental value of 2.50 Å [18,19], and with the value of 2.51 Å obtained from DFT calculations for monolayer *h*-BN [20,21]. In addition, the cohesive energy per atom was determined using the expression $E_{coh} = \left(E^{\text{BN}}_{\text{pair}} - E^{\text{B}} - E^{\text{N}}\right)/\text{N}$, where $E^{\text{BN}}_{\text{pair}}$ is the total energy per pair of B-N atoms in the equilibrium structure. The terms E^{B} and E^{N} denote the total energies of the free B and N atoms respectively. The cohesive energy of -14.6 eV (vdW-DF), -11.88 eV (GGA-PBE) and -9.36 eV (LDA) was obtained for pristine *h*-BN bilayer. Table 2 shows the stereo-electronic properties of *h*-BN bilayer from the three XC functional. The vdW-DF predicts the strongest structural cohesion compared to GGA-PBE and LDA. The strong binding from vdW-DF is also seen in the short interlayer separation (3.32 Å) compared to 4.28 Å obtained from GGA-PBE. The interlayer distance is ~0.9% shorter than the experimental interlayer distance (3.35 Å) expected in bilayer *h*-BN. These indicate that dispersive interactions play a nontrivial role in the determination of the cohesive properties of bilayer *h*-BN.

One way of obtaining a defective *h*-BN bilayer is through the application of uniaxial strain. The strain, in this case, is applied along the *c*-axis (*i.e.* normal to the stacking direction). Figure 2 shows the effect of uniaxial strain on the electronic band edges. The top (and bottom) panels in Fig. 2

correspond to decreasing (and increasing) interlayer distances respectively. We find that the shift in energy between the two top-most lying π-bands at points *K* (valence band) and *K'* (conduction band) increases as the interlayer separation is reduced from the value in the optimised structure. The corresponding positions of the HOMO and LUMO states also respond to applied uniaxial compression, and give rise to decreased band gaps. It is noted that the widest band gap occurs at the equilibrium interlayer separation of 3.32 Å. Taken together, the increase in band gap with increasing interlayer distances up to the equilibrium interlayer separation is ascribed to the strain-induced removal of the π-band degeneracy at point *K* - suggesting the possibility of tuning the band gap in bilayer *h*-BN by strain engineering. The lower panels of Fig. 2 show that application of uniaxial tensile strain does not shift the energy levels of the π-bands. Thus, no change is observed in the band gap, although the application of uniaxial tensile strain removes the degeneracy of the bands.

Our vdW-DF band structure calculations show increased in-plane band distortions as interlayer distance is decreased. These indicate that electrons acquire substantially higher speeds within the plane. Out-of-plane band distortions, i.e. along Γ→A, are minimal even at very small interlayer distances for which the van der Waals interactions are strong. At large interlayer distances, therefore, we cannot attribute the observed crossings in the three lower conduction bands to enhanced out-of-plane electron mobility due to substantially weakened van der Waals interactions. In addition, because the band dispersions are correctly described within the plane of the bilayer (i.e. Γ→M→K→Γ) at short interlayer distances, we attribute the non-monotonous variation of the Γ-point energy, observed at 6, 7 and 8 Å, to the decoupling of the bilayer at large distances.

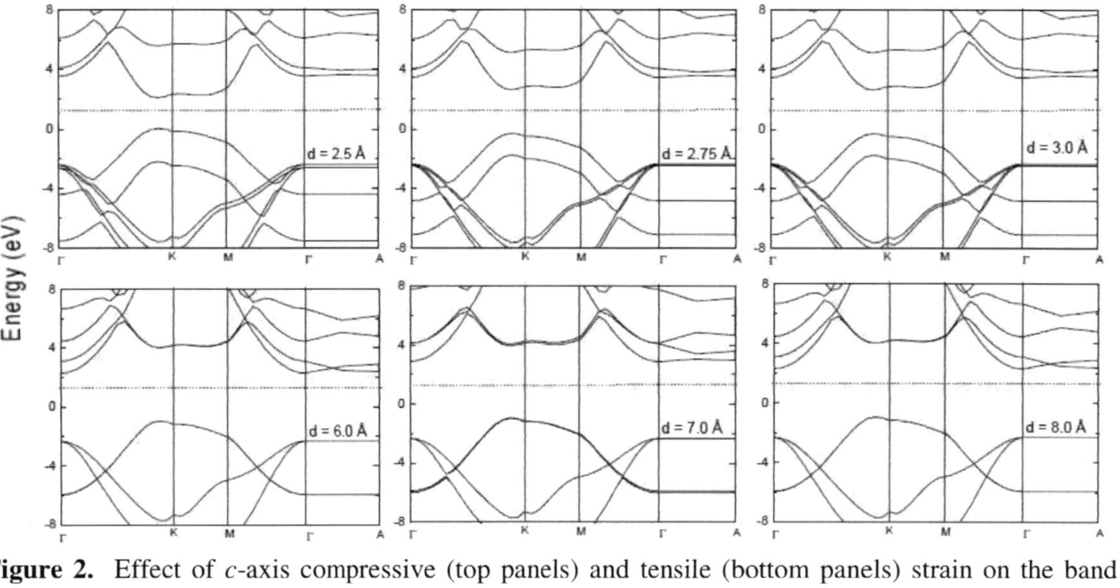

Figure 2. Effect of *c*-axis compressive (top panels) and tensile (bottom panels) strain on the band edges for varying interlayer separations. The Fermi level is denoted with horizontal lines.

Reduced band gaps can also obtained by breaking the symmetry of the ordered honeycomb structure. However, only the intrinsic defects, shown in Fig. 3 (a)-(i), are considered in the present study. The total energy was also converged with respect to supercell size. The difference ΔE between the total energies of the defect-free and defective structures of size 3×3 and 4×4 converges to 31 meV for V_B+V_B complex. Because this configuration has the most extended open-volume, we conclude that the total energy of the defective system is well converged. A 3×3 supercell with lattice parameter $a = b$ = 7.53 Å was used in the calculations reported here. In the present study, the local structure of the complexes is not constrained to nearest neighbour positions unlike in our previous studies of defects in monolayer *h*-BN [22]. Stoichiometric defects that can form on one layer, or on both layers, of the bilayer structure are denoted with an asterisk. We note that their total energies are lower when they form on one layer compared to both layers. When formed in one layer in each case, the total energy of the V_N+V_B complex is 1.47 eV lower while the Stone-Wales defect is 8.71 eV lower than in both layers. The formation of nonstoichiometric complexes (as point defects on both layers) give total energies that are at least 20.41 eV lower than when the defects are formed on only one layer.

Fig. 3(j) shows the formation energy of neutral defects as a function of changes in chemical potential, $\Delta\mu$. The formation energies are constrained to vary within the interval, $-\Delta H_f \leq \Delta\mu \leq +\Delta H_f$, fixed by the formation energy ΔH_f of the pristine bilayer. We calculated ΔH_f using $\Delta H_f = -(\mu_{BN} - \mu_B^{bulk} - \mu_N^{bulk})$, and obtained the value of -2.58 eV in good agreement with the experimental value of -2.60 ± 0.02 eV [23]. The boron-rich and nitrogen-rich conditions correspond to limits $+\Delta H_f$ and $-\Delta H_f$ respectively. The formation energy of nonstoichiometric defects is dependent on $\Delta\mu$, as expected. The V_N and its complexes are more stable than the V_B and its complexes in B-rich condition, and vice versa. The creation of the double N (or B) antisite complex at B (or N) sites on both layers, or on nearest neighbour sites on one of the layers presents the lowest formation energy in N-rich and B-rich conditions, respectively. This trend is also seen in the double antisite complexes of B (N_B+N_B) and N (B_N+B_N) when they are created on both top and bottom layers. We find that nonstoichiometric complexes that induce excess nitrogen or boron, in each case, are relatively more stable in the growth condition that is rich in the excess atomic species.

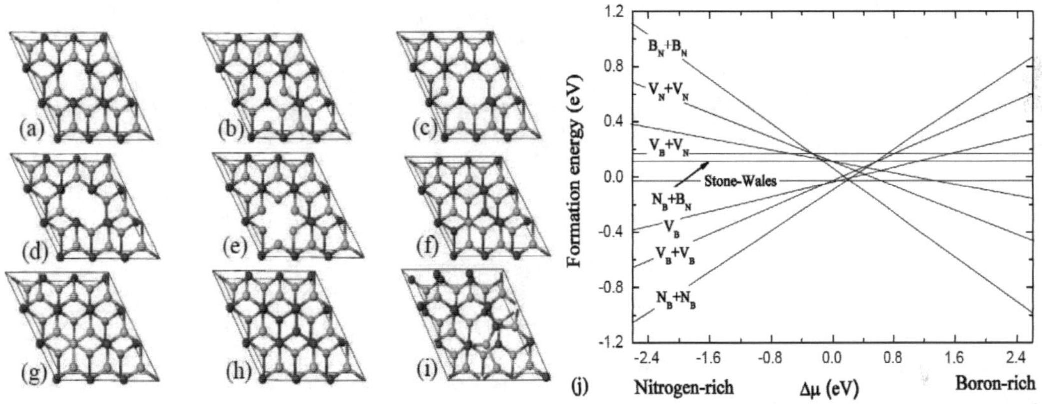

Figure 3. (a) V_B, (b) V_N, (c) V_N+V_B(*), (d) V_B+V_B, (e) V_N+V_N, (f) N_B+N_B, (g) B_N+B_N, (h) N_B+N_B, (i) Stone-Wales (SW) defect (*), and (j) their energies of formation as a function of changes in atomic chemical potential.

The electronic density of states (EDOS) gives an estimate of how closely packed the energy levels are distributed within an energy interval. Because the band structure is modified by defects, the defect states are required to form *e*-resonances with the top of the valence band (VB) or bottom of the conduction band (CB) in order to maintain the semiconducting band gap. When a defect state forms an *s*-resonance at the Fermi level, it eliminates the band gap by bridging the VB and CB. This induces metallic transport properties. We find that boron and nitrogen vacancies and their complexes introduce *s*-resonance to the Fermi level in bilayer *h*-BN. Figure 4 shows the EDOS of the V_B+V_N complex and the Stone-Wales (SW) defect. Both defects are formed on one layer. The Fermi levels correspond to -0.38 eV (V_B+V_N) and 0.31 eV (SW) respectively. All other defects introduce defect states that form *s*-resonances with the Fermi level. In the monolayer for instance, replacing a boron atom with the carbon impurity (C_B) leads to the highest stability [22,23]. More importantly, because the smallest band gap is obtained in C_B [24], we suggest the possibility of further reducing the band gaps in V_B+V_N and SW layers by impurity adsorption.

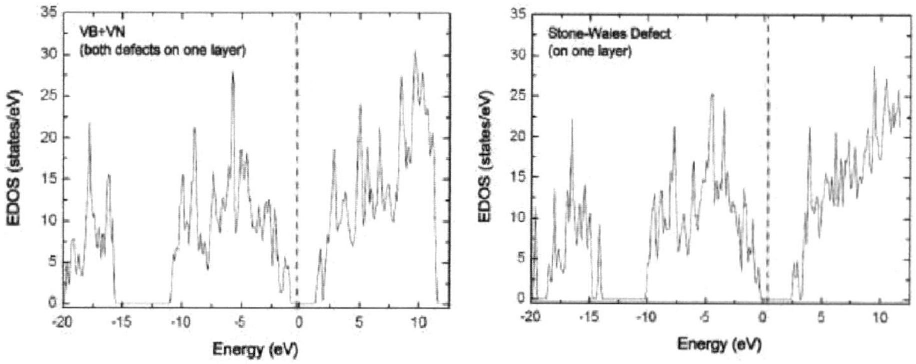

Figure 4. Electronic density of states in V_B+V_N complex and the Stone-Wales defect.

4. Conclusion

A comparative density functional study of the ground state properties of bilayer *h*-BN has been presented. We conclude that the vdW-DF gives a better description of the cohesive properties while the GGA-PBE gives the widest band gap in agreement with experiments. Using the vdW-DF, it is shown that band gap modification in the *h*-BN bilayer is achievable by application of uniaxial compressive strain, and by introduction of defects. Only the stoichiometric V_B+V_N complex and the Stone-Wales defect are able to preserve the band gap. Both defects present their lowest formation energies when they form on either the top or bottom layer of the coupled *h*-BN bilayer.

Acknowledgements

This work was supported by the University of Pretoria under E2020 Project No. 5. NC is grateful to the National Institute for Theoretical Physics for support.

References

[1] Gorbachev R V, Riaz I, Nair R R, *et al.* 2011 *Small* 7 465
[2] Young A F, Dean C R, Meric I, *et al.* 2010 *arXiv:1004.5556*
[3] Dean C R, Young A F, Meric I, *et al.* 2010 *Nature Nanotech.* **5** 722
[4] Lee C, Li Q, Kalb W, *et al.* 2010 *Science* **328** 76
[5] Ponomarenko L A, Yang R, Mohiuddin T M, *et al.* 2009 *Phys. Rev. Lett.* **102** 206603
[6] Golberg D, Bando Y, Huang Y, *et al.* 2010 *ACSNANO* **4** 2979
[7] Giannozzi P *et al.* 2009 *J. Phys Condens. Matter* **21** 395502
[8] Perdew J P and Zunger A 1981 *Phys. Rev.* B **23** 5048
[9] Perdew J P, Burke K and Ernzerhof M 1996 *Phys. Rev. Lett.* **77** 18
[10] Barone V, Casarin M, Forrer D, *et al.* 2009 *J. Comput. Chem.* **30** 934
[11] Dion M, Rydberg H, Schroder E, *et al.* 2004 *Phys. Rev. Lett.* **92** 246401
[12] Roman-Perez G and Soler J M 2009 *Phys. Rev. Lett.* **103** 096102
[13] Kresse G and Joubert D 1999 *Phys. Rev.* B **59** 1758
[14] Blöchl P E 1994 *Phys. Rev.* B **50** 17953
[15] Monkhorst H J and Pack J D 1976 *Phys. Rev.* B **13** 5188
[16] Ribeiro R M and Peres N M R 2011 *Phys. Rev.* B **83** 235312
[17] Song L, Ci L, Lu H, *et al.* 2010 *Nano Lett.* **10** 3209
[18] Meyer J C, Chuvilin A, Algara-Siller G, *et al.* 2009 *Nano Lett.* **9** 2683
[19] Li J, Gui G and Zhong J 2008 *J. App. Phys.* **104** 094311
[20] Azevedo S, Kaschny J R, de Castilho C M C, *et al.* 2009 *Eur. Phys. J.* B **67** 507
[21] Solozenko V L, Will G and Elf F 1995 *Solid State Commun.* **96** 1
[22] Ngwenya T B, Ukpong A M and Chetty N 2011 *Phys. Rev.* B **84** 245425
[23] Wise S S, Margrave J L, Feder H M, Hubbard W N 1966 *J. Phys. Chem.* **70** 7
[24] Jin C, Lin F, Suenaga K and Iijima S 2009 *Phys. Rev. Lett.* **102** 195505

Journal of Physics: Conference Series **367** (2012) 012005 doi:10.1088/1742-6596/367/1/012005

Non-linear electro-elastic coupling in non-centrosymmetric materials

J Even[1], L Pedesseau[1], C Hajlaoui[1], C Katan[1,2] and J-M Jancu[1]
[1] Université Européenne de Bretagne, INSA, FOTON, UMR 6082, F-35708 Rennes, France
[2] CNRS, Institut des Sciences Chimiques de Rennes, UMR 6226, 35042 Rennes, France

E-mail: jacky.even@insa-rennes.fr

Abstract. We propose an extended symmetry-based thermodynamical model of third-order electro-elastic coupling. The entanglement of electrostriction non-linear piezoelectricity and other non-linear phenomena, is studied in details at the same level of theory. Symmetry properties of materials and nanostructures are taken into account. Density Functional Theory (DFT) is used to calculate the complete set of linear and non-linear coefficients. Electrostriction dominates for nitride compounds in the Würtzite structure, but non-linear elasticity and piezoelectricity must be taken into account for strain and electric field evaluation.

1. Introduction
A complete model of third-order electro-elastic coupling has recently been proposed [1]. It is an extension of a previous thermodynamical approach [2] (quoted as ATK model in this paper). The latter seems to be almost ignored up to now, although a number of misinterpretation or errors could have been avoided recently in the literature. In this work, the ATK model is combined systematically with symmetry analyses and DFT calculations to evaluate consistently the various linear and non-linear coefficients, especially in non centrosymmetric materials. Symmetry properties of third-order elastic constants are well known [3], but have been extensively studied for third-order coupled constants only more recently [4]. In this paper, third-order coefficients are mostly determined from finite difference studies of material's polarisation or stress tensor under various electrical or strain conditions. We show in section 2. that the non-linear tensors can be strongly entangled. Actually, in the same theoretical model, coefficients defined for a specific thermodynamical potential should not be mixed with others without caution. Experimental determination of piezoelectric and electrostrictive entangled components must also rely on careful definitions of experimental conditions which makes such measurements very tedious and tricky [5]. The direction of the applied field and the frequency behaviour are useful tools to separate the strains which are arising from piezoelectricity and electrostriction.

2. Thermodynamic potentials and derivatives

2.1. Thermodynamic potentials and first order derivatives
Thermodynamic potentials are introduced using standard definitions for the free energy and enthalpy, $F = U - TS$ and $G = U - \sigma_i \eta_i - TS$, and the transformed free energy and enthalpy

$F_e = U - TS - E_l D_l$ and $G_e = U - \sigma_i \eta_i - TS - E_l D_l$. The D_l, E_l vectors and the η_i, σ_i strain-stress tensors (in Voigt notation) are associated respectively to the l, m, n, u, v, o and i, j, k, p, q, r indices varying between 1-3 and 1-6 respectively.
The first order derivatives of the thermodynamic potentials are:

- $dG = -SdT - \eta_i d\sigma_i + E_l dD_l$
- $dG_e = -SdT - \eta_i d\sigma_i - D_l dE_l$
- $dF_e = -SdT + \sigma_i d\eta_i - D_l dE_l$
- $dF = -SdT + \sigma_i d\eta_i + E_l dD_l$

In the following, most expressions are related to F_e, G and G_e.

2.2. Second order derivatives

Three second-order derivatives in relation to D_l, E_l, η_i and σ_i are defined for each of the thermodynamic potentials.

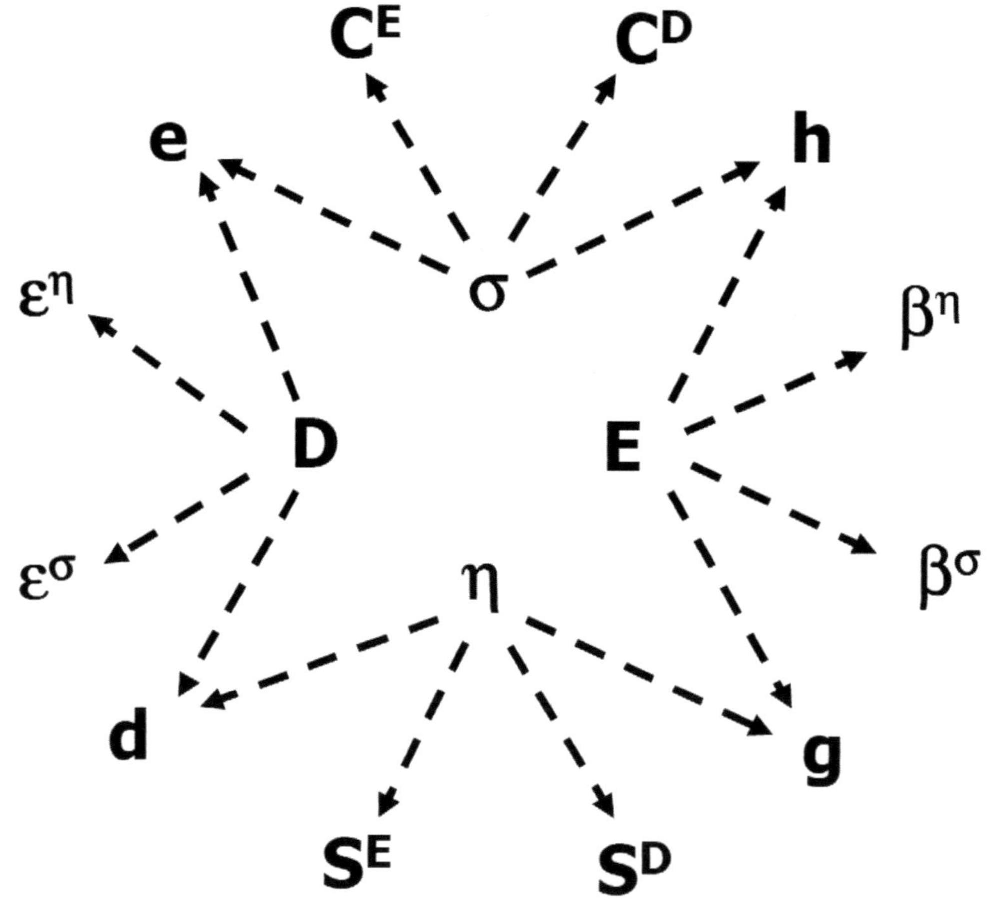

Figure 1. Schematic representation of the relations between thermodynamic variables (inner variables) and second order derivatives of the thermodynamic potentials (outer variables). For clarity, minus signs are not shown.

Figure 1 is a schematic representation of the relations between thermodynamic variables and second order derivatives of the thermodynamic potentials (for the clarity of the figure minus signs are not shown) :

- $d\sigma_i = C^E{}_{ij} d\eta_j - e_{il} dE_l$ and $dD_l = e_{li} d\eta_i + \varepsilon^\eta{}_{lm} dE_m$ for F_e
- $d\eta_i = S^E{}_{ij} d\sigma_j + d_{il} dE_l$ and $dD_l = d_{li} d\sigma_i + \varepsilon^\sigma{}_{lm} dE_m$ for G_e
- $d\eta_i = S^D{}_{ij} d\sigma_j + g_{il} dD_l$ and $dE_l = -g_{li} d\sigma_i + \beta^\sigma{}_{lm} dD_m$ for G

The second-order derivatives of the thermodynamic potentials are related to each other by exact expressions:

- $\varepsilon^\sigma{}_{lm} - \varepsilon^\eta{}_{lm} = e_{lj} S^E{}_{ji} e_{im}$
- $e_{il} = C^E{}_{ij} d_{jl}$
- $S^E{}_{ij} - S^D{}_{ij} = g_{il} d_{lj}$

2.3. Third order derivatives

Four third-order derivatives are defined for each of the thermodynamic potentials. Figure 2 is a schematic representation of the relations between thermodynamic variables and third order derivatives of the thermodynamic potentials (minus signs are not shown). Some of these derivatives are common to different thermodynamic variables, for example :

- $B_{lij} = \left.\frac{\partial e_{li}}{\partial \eta_j}\right|_{E,T} = B_{ijl} = -\left.\frac{\partial C_{ij}}{\partial E_l}\right|_{\eta,T} = -\left.\frac{\partial^2 \sigma_i}{\partial \eta_j \partial E_l}\right| = \left.\frac{\partial^2 D_l}{\partial \eta_i \partial \eta_j}\right|_{E,T} = -\left.\frac{\partial^3 F_e}{\partial \eta_i \partial \eta_j \partial E_l}\right|$

The third order derivatives are also connected to second order derivatives :

- $dC^E{}_{ij} = C^E{}_{ijk} d\eta_k - B_{ijl} dE_l$, $de_{li} = B_{lij} d\eta_j + L_{\lim} dE_m$ and $d\varepsilon^\eta{}_{lm} = L_{lmi} d\eta_i + \varepsilon^\eta{}_{lmn} dE_n$
- $dS^E{}_{ij} = S^E{}_{ijk} d\sigma_k + F_{ijl} dE_l$, $dd_{li} = F_{lij} d\sigma_j + M_{\lim} dE_m$ and $d\varepsilon^\sigma{}_{lm} = M_{lmi} d\sigma_i + \varepsilon^\sigma{}_{lmn} dE_n$
- $dS^D{}_{ij} = S^D{}_{ijk} d\sigma_k + R_{ijl} dD_l$, $dg_{li} = R_{lij} d\sigma_j + Q_{lmi} dD_m$ and $d\beta^\sigma{}_{lm} = -Q_{lmi} d\sigma_i + \beta^\sigma{}_{lmn} dD_n$

Most of these quantities are already used in various papers [5-7], but we will focus on the non-linear piezoelectric tensors B_{lij}, F_{lij}, R_{lij} and the electrostrictive tensors L_{lmi}, M_{lmi} and Q_{lmi}. The third-order derivatives of the thermodynamic potentials are also related to each other by various exact expressions:

- $F_{lij} = S^E{}_{ik} B_{lkp} S^E{}_{pj} + e_{lk} S^E{}_{kij}$
- $R_{lij} = \beta^\sigma{}_{lm} F_{mij} - \beta^\sigma{}_{lm} \beta^\sigma{}_{nv} (M_{mni} d_{vj} + M_{mnj} d_{vi}) - \beta^\sigma{}_{lmn} d_{mi} d_{nj}$
- $Q_{lmi} = \beta^\sigma{}_{lmn} d_{ni} + \beta^\sigma{}_{mn} \beta^\sigma{}_{lv} M_{vni}$
- $M_{lmi} = (B_{lkp} d_{mp} + B_{mkp} d_{lp}) S^E{}_{ki} + L_{lmj} S^E{}_{ji} + e_{lk} e_{mj} S^E{}_{kji}$.

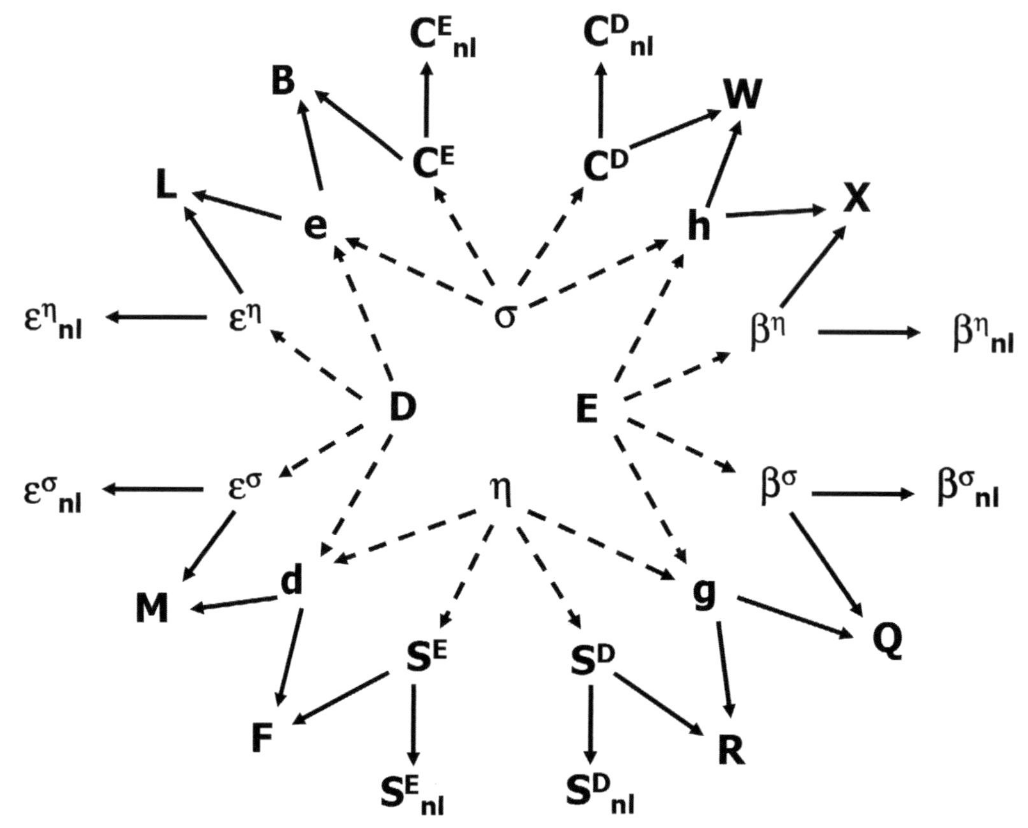

Figure 2. Schematic representation of the relations between thermodynamic variables, second derivatives and third order derivatives (outer variables) of the thermodynamic potentials. For clarity, minus signs are not shown.

3. Symmetry analysis

It has been shown in section 2. that the non-linear tensors can be strongly entangled. For instance, in the case of non-centrosymmetric materials, an additional term related to the linear piezoelectricity and the non-linear susceptibility must be added to the commonly used relation $Q_{lmi} = \beta^{\sigma}{}_{mn}\beta^{\sigma}{}_{lv}M_{vni}$ [5]. It is necessary to use symmetry properties of the materials in order to check whether or not the entanglement has to be taken into account. In order to get a first insight into the influence of lattice symmetry, we may compare materials which crystallise in a cubic lattice associated either to the m3m (e.g. diamond or undistorted perovskite lattices) and 432 point groups. In both cases, the linear piezoelectric tensors and non-linear susceptibilities are vanishing: $e_{li} = d_{li} = g_{li} = 0$ and $\varepsilon^{\eta}{}_{lmn} = \varepsilon^{\sigma}{}_{lmn} = \beta^{\sigma}{}_{lmn} = 0$. Only six independent non-linear elastic constants and compliances exist: $C^{E}{}_{ijk}$, $S^{E}{}_{ijk} = S^{D}{}_{ijk}$. Q_{lmi}, M_{lmi} and L_{lmi} tensors related to electrostriction are simply connected, like in isotropic materials: $Q_{lmi} = \beta^{\sigma}{}_{mn}\beta^{\sigma}{}_{lv}M_{vni}$ and $M_{lmi} = L_{lmj}S^{E}{}_{ji}$, although the number of independent coefficients is larger. The only difference between the m3m and 432 point groups comes from the non-linear piezoelectricity phenomenon which appears only in the 432 case. Among the thirty-two crystal classes, eleven are centrosymmetric (m3m case) and twenty-one are non-centrosymmetric, but of these, 432 is a special case. For point group 432, non-linear piezoelectricity

Journal of Physics: Conference Series **367** (2012) 012005 doi:10.1088/1742-6596/367/1/012005

and electrostriction, are indeed disentangled. The various non-linear piezoelectric tensors are simply connected and only one independent component B_{124} has to be determined: $F_{124} = (S^E_{11} - S^E_{12})B_{124}S^E_{44} = \varepsilon^{\sigma} R_{124}$. Ten of the non-centrosymmetric crystal classes represent the polar crystal classes, which show a spontaneous polarization. For all this crystal classes, the entanglement of non linear piezoelectricity, non linear dielectric susceptibility and electrostriction is very strong. The hexagonal 6mm polar crystal class corresponds to a number of important materials in the wurtzite structure. The tensors describing non-linear piezoelectricity have 8 independent components : B_{311}, B_{312}, B_{313}, B_{333}, B_{344}, B_{115}, B_{125}, B_{135} whereas the tensor describing electrostriction has 6 components : L_{11}, L_{12}, L_{13}, L_{31}, L_{33}, L_{44}. Using relation $M_{lmi} = (B_{lkp}d_{mp} + B_{mkp}d_{lp})S^E{}_{ki} + L_{lmj}S^E{}_{ji} + e_{lk}e_{mj}S^E{}_{kji}$, we can compare the non-linear piezoelectric, electrostrictive and elastic contributions. In the case of nitride compounds like GaN, the electrostrictive contribution is the most important, but non-linear piezoelectric effect can not be ignored [1].

4. Determination of non-linear coefficients using density functional theory

The methods of density-functional perturbation theory (DFPT) may be used to calculate various physical responses. In fact the efficient use of the "2n+1" theorem [8], using only by-products of a first-order perturbation calculation, in principle gives the second and third-order derivatives of the total energy at the level of the ATK model, if the atomic-displacement variables are eliminated. Second-order derivatives may be used with an existing DFPT implementation [9] to calculate various physical response properties of insulating crystals, including elastic constants, linear piezoelectric tensors, linear dielectric susceptibility, as well as tensor properties related to internal atomic displacements like Born charges. Third-order derivatives of the total energy are related to a number of physical properties described in this work, like the non-linear electrical susceptibility, the non-linear elasticity, the photo-elastic and electrostrictive effects. However, most practical implementations of the DFPT are restricted to some quantities related to internal atomic displacements. DFT simulations are performed using a state of the art DFT implementation [9] within the local density approximation (LDA). Plane wave basis sets were used with a kinetic energy cutoff of 950 eV after convergence studies. The sets of k-points were generated following the procedure of Pack and Monkhorst and Pack [10] namely 10 10 10. Polarisation and stress tensor components can be studied as a function of the electric field and strain tensor components. We applied various electrical or strain conditions for the DFT determination of all the coefficients in the case of the GaN and AlN crystals [1]. In figure 3, the variations of σ_3 and P_3 are reported as a function of η_3 for AlN.

Journal of Physics: Conference Series **367** (2012) 012005 doi:10.1088/1742-6596/367/1/012005

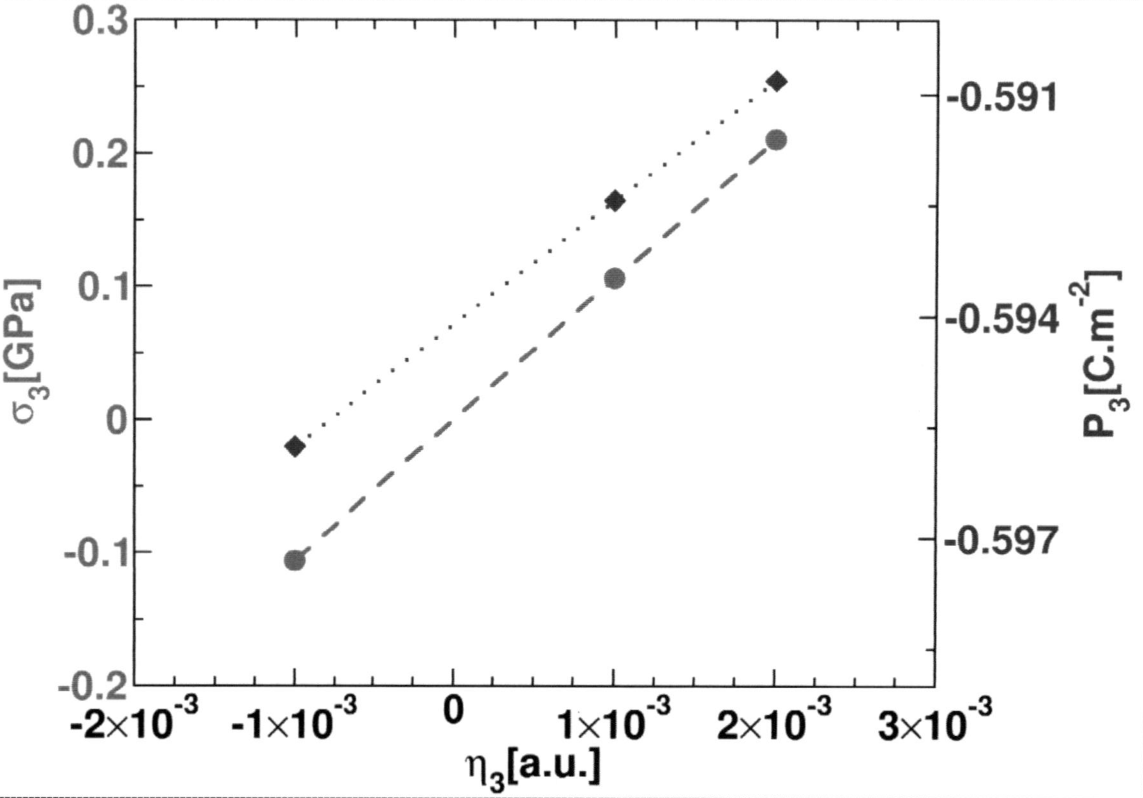

Figure 3. Variations of σ_3 and P_3 calculated using a state of the art DFT implementation [9] are reported as a function of η_3 for AlN.

Acknowledgement.
This work was performed using HPC resources from GENCI-CINES (Grant 2010-[c2010096472]) and French ANR Project INSCOOP (no ANR-11-NANO-012)

References

[1] Pedesseau L, Katan C and Even J 2012 *Appl. Phys. Lett.* **100** 031903
[2] Adam W, Tichy J and Kittinger E 1988 *J. Appl. Phys.* **84** 2556
[3] Brugger K 1965 *J. Appl. Phys.* **36** 759
[4] Grimmer H 2007 *Acta Cryst* **A63** 441
[5] Newnham R E, Sundar V, Yimnirun R, Su J and Zhang Q M 1997 *J. Phys. Chem.* **B101** 10141
[6] Bester G, Wu X, Vanderbilt D and Zunger A 2006 *Phys. Rev. Lett.* **96** 187602
[7] Gerlich D and Breazeale M A 1990 *J. Appl. Phys.* **68** 5119
[8] Hamann D R, Wu X, Rabe K M and Vanderbilt D 2005 *Phys. Rev.* **B71** 035117
[9] Gonze X, Amadon B, Anglade P M, Beuken J M, Bottin F, Boulanger P, Bruneval F, Caliste D, Caracas R, Cote M, Deutsch T, Genovese L, Ghosez Ph, Giantomassi M, Goedecker S, Hamann D R, Hermet P, Jollet F, Jomard G, Leroux S, Mancini M, Mazevet S, Oliveira M J T, Onida G, Pouillon Y, Rangel T, Rignanese G M, Sangalli D, Shaltaf R, Torrent M, Verstraete M J, Zerah G, Zwanziger J W, 2009 *Computer Phys. Commun.* **180** 2582
[10] Monkhorst H J and Pack J D, 1976 *Phys. Rev.* **B13**, 5188

Journal of Physics: Conference Series **367** (2012) 012006 doi:10.1088/1742-6596/367/1/012006

Strain dependence of Internal Displacement and Effective Charge in Wurtzite III-N semiconductors

Joydeep Pal[1]*, Geoffrey Tse[1], Vesel Haxha[1], Max A. Migliorato[1] and Stanko Tomić[2]

[1]School of Electrical and Electronic Engineering, University of Manchester, United Kingdom

[2]Joule Physics Laboratory, School of Computing, Sciences and Engineering, University of Salford, United Kingdom

* joydeep.pal@postgrad.manchester.ac.uk

Abstract. The elastic and dielectric properties of binary III-N wurtzite semiconductors have been investigated as a function of strain. Using an ab initio density functional theory (DFT), we concentrate on the internal displacement (u) and Born effective charge (Z*) and show that our model provides a unique non linear dependence of the III-N material properties as a function of strain.

1. Introduction

Group III-V semiconductors are gaining popularity by virtue of the wide range of prospective applications as optoelectronic devices such as lasers and LEDs along with electronic devices[1] such as HEMTs, HBTs, Gunn diodes, transducers and micropositioners. III-Nitrides can be employed in high frequency and high power devices for radar and avionics exploiting material properties such as their wide bandgap and high electron mobility.[2] Strain induced electrical and optical along with electromechanical properties of the III-N materials are well known to strongly affect device behaviour.[3,4,5] Substantial nonlinear effects in the strain to second order have been reported already in wurtzite III-N[6,7,8,9,10,11] following previous analogous reports for zincblende InGaAs.[12,13]

2. The Physical Model And DFT Calculations

Our previous work[6] has already reported the methodology for computing the piezoelectric coefficients (PZCs). Such methodology is based on a semi empirical formalism within Density Functional Theory (DFT), Local Density Approximation (LDA) and plane wave pseudopotentials. The same approach has been exploited previously to obtain predictions for InGaAs which seem to provide excellent agreement with experimental data.[12,14] The piezoelectric polarization is calculated as the sum of a direct dipole contribution and a bond contribution, a semi empirical approach, originally proposed by Harrison[15] :

$$P_{\hat{x}_i} = \frac{Z_H^* \delta r + 2\alpha_p \left(1-\alpha_p^{\,2}\right) \cdot \sum_{q=1}^{4} \left(\vec{r}_q \cdot \hat{x}_i\right) \delta R_q}{2\Omega} \qquad \text{Eq 1}$$

In Eq.1, $\hat{x}_i$ is the direction of the Cartesian axes, δr is the displacement vector of cations in respect of anions from the initial position (i.e. the condition when all the bonds in the tetrahedron are equal in nature), r_q and δR_q are the distance and displacement (deviation from the initial position) vectors of the nearest neighbour q from the atom at the centre of the tetrahedron, respectively, α_p is the bond polarity and Ω is the atomic volume, Z_H^* is the atomic charge, not to be confused with the transverse effective charge or Born charge (Z*), calculated with DFPT. Our previous work on wurtzite III-N[6] and zincblende GaAs and InAs[12,14] has already reported that the value of the atomic charge Z_H^* needs to be 25% of the value for the dynamic Born effective charge (Z*) to attain precise estimates of piezoelectric polarization compared to the experimental values. However Z* is linked to the bond polarity α_p so it is still an important quantity.

In this work, we report the non linear behaviors of two quantities, the deformation or internal displacement parameter (u) and Z*, which are fundamental for calculating Eq.1. These have been calculated for both bulk and strained cases using pseudopotentials derived with the Troullier–Martin scheme (for u),[16] along with the Hamann scheme (for Z*),[17] in the CASTEP[18] code.

Both Density functional theory in the local density approximation (DFT-LDA)[19] and density functional perturbation theory (DFPT) have been used in the calculations. Single-particle orbitals expressed in a plane-wave basis set with kinetic energy of up to 1keV and Brillouin zone summations of up to 10x10x6 Monkhorst-Pack k-point grids[20] were sufficient to make the simulations converge below a remaining error of about 1%. Z* was computed using the Berry phase approach[21] with application of periodic boundary conditions and finite electric field perturbation.

3. Results

Our previous works reported the DFT calculated equilibrium values for the lattice parameters, internal displacement parameter (u), Z* and the resulting bond polarity α_p (obtained from Z*) for all three III-N semiconductors and in Table I we showed these to be in good agreement with the existing values in the literature.[22,23,24,25]

Table I: Comparison of the physical parameters of Group-III Nitrides (GaN, AlN and InN) used in this work[6] with other calculated values in brackets.

Parameters	GaN	AlN	InN
a (Å)	3.155 (3.196)[22] (3.131)[23]	3.063 (3.076)[22] (3.069)[23]	3.523 (3.524)[22] (3.509)[23]
c (Å)	5.149 (5.220)[22] (5.104)[23]	4.906 (4.980)[22] (4.910)[23]	5.725 (5.773)[22] (5.656)[23]
u (Å)	0.376 (0.376)[22,23] (0.377)[24]	0.382 (0.380)[22] (0.382)[23,24]	0.377 (0.377)[22,24] (0.379)[23]
Z*	2.583 (2.510)[23]	2.553 (2.652)[23]	2.850 (3.045)[23]
α_p	0.517 (0.511)[25]	0.511 (0.508)[25]	0.578 (0.542)[25]

In Fig. 1 we show the distribution of the dimensionless internal displacement parameter (u) with parallel and perpendicular strain (varying from -10% to 10%).

The results show a wide variation of the u-parameter. Different combinations of strain applied to the system leads to different values for AlN, GaN and InN, but essentially the same behaviour.

The range of the internal displacement parameter with applied strain differs with different materials. While for both GaN and InN, the u-parameter dips to 0.34 for 10% strain along c-axis and to 0.35 for AlN. Application of strain parallel to growth direction reveals different trends altogether, where the u-parameter for AlN and InN grows to 0.44 and for GaN to 0.43 for 10% strain.

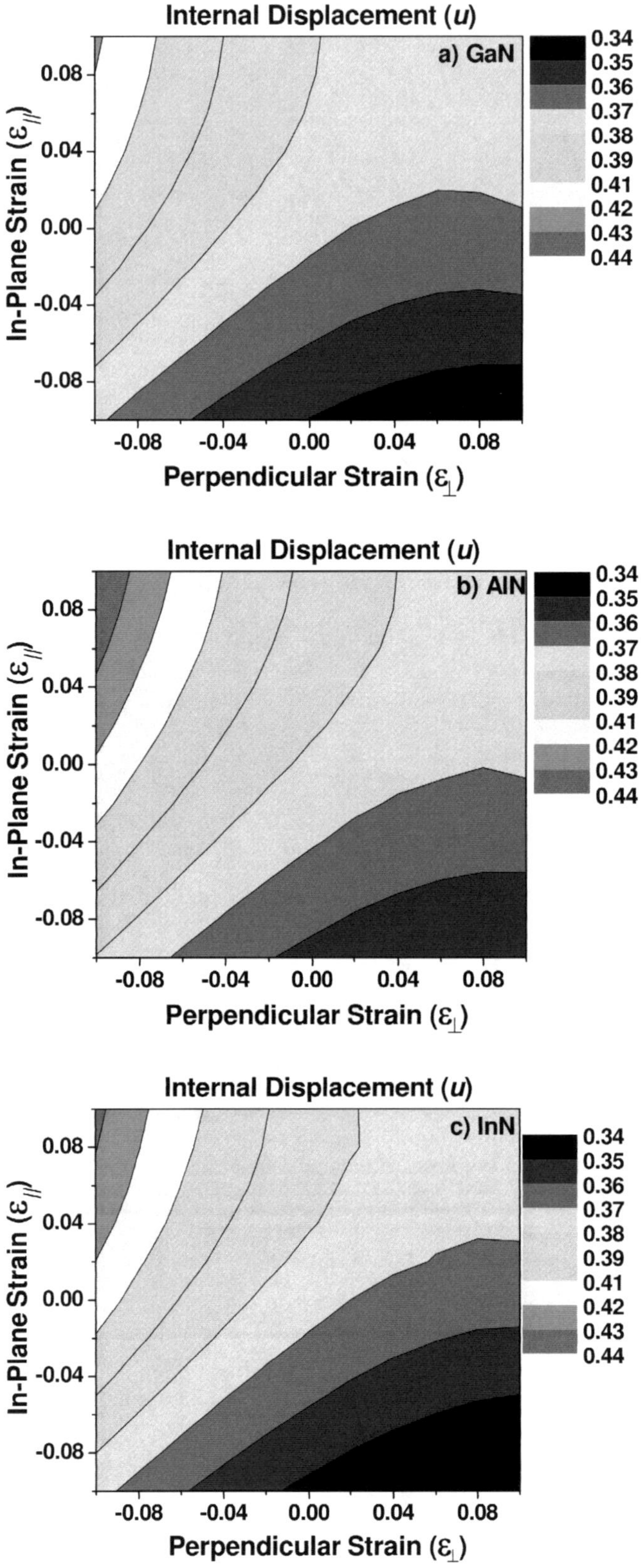

Fig.1: Strain dependence of internal displacement parameter (u) as a function of in-plane and perpendicular strain (from -0.1 to 0.1 in steps of 0.02) for a) GaN, b) AlN and c) InN respectively.

The u-parameter is dimensionless.

Journal of Physics: Conference Series **367** (2012) 012006 doi:10.1088/1742-6596/367/1/012006

We also investigated the strain dependence of Z* for all III-N materials for various arrangements of strain along in-plane and perpendicular to the growth plane directions. The outcomes of the Z* calculations reflect an unusual trend of non linearity on strain which again is common to GaN, AlN and InN.

In Fig.2, the strain effects on Z* for strain along perpendicular as well as in-plane strain varying ±10% with steps of 2% are shown. While GaN and AlN show almost identical behaviour, InN shows a less pronounced non linear dependence on perpendicular strain. But Z* for all the three different III-Nitride materials, namely GaN, AlN and InN, increases with more tensile strain in the plane of the direction of growth.

We have not explored the non linear dependence of u and Z* on shear strain in the growth direction or growth plane. It could have a significant influence on nanostructures such as quantum dots, but not on the bulk semiconductors or 2-dimensional thin films, which has been the primary aim of this work.

4. Conclusions

Our semi empirical formalism has been employed to investigate the effect on fundamental and critical elements of III-N materials. We have already illustrated non linear effects on piezoelectric coefficients in our previous work[6] leading towards new and improved sets of parameters for III-N binary and ternary materials. In this work, we have shown that a strong strain effect of second-order is prevalent on material properties such as internal displacement (u) and born effective charge (Z*). As this data was obtained in the framework of DFT-LDA, it can be useful in fitting other semiempirical model such as force fields for Molecular Dynamics simulations.

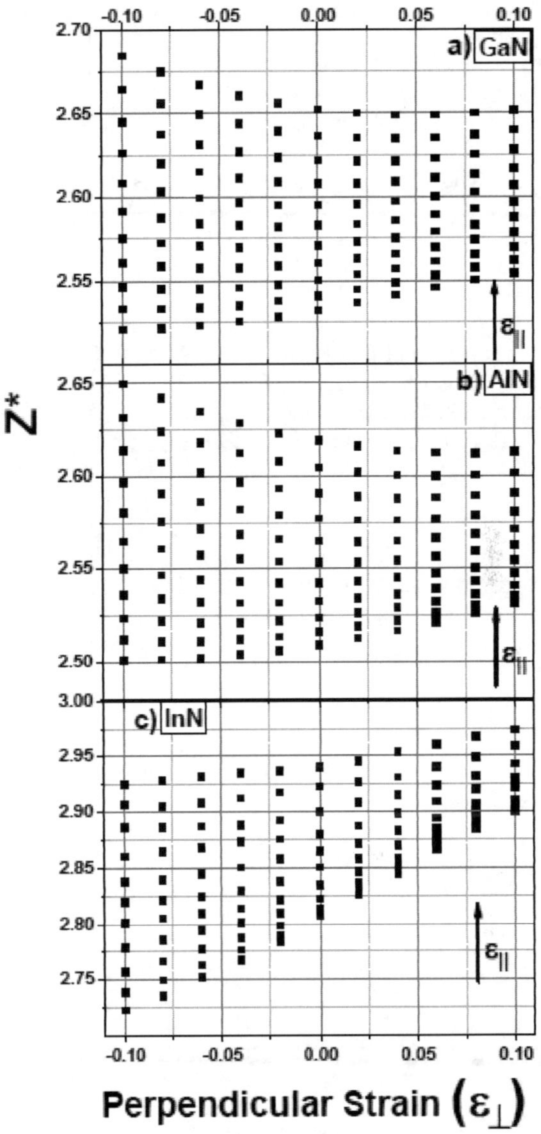

Fig.2: Z*(expressed in terms of the electron charge **e**) as a function of perpendicular strain (from -0.1 to 0.1 in steps of 0.02) along with in-plane strain, following similar steps as perpendicular strain, for a) GaN, b) AlN and c) InN respectively.

References

[1] S. Nakamura, G. Fasol, The Blue Laser Diode: GaN Based Light Emitters and Lasers (Springer-Verlag, Berlin, 1997)

[2] W. D. Hu, X. S. Chen, F. Yin, J. B. Zhang, and W. Lu, J. Appl. Phys, **105**, 084502 (2009).

[3] A. D. Andreev and E. P. O'Reilly, Phys Rev B **62**, 15851 (2000)

[4] S. Tomić and N. Vukmirović, Phys. Rev. B **79**, 245330 (2009)
[5] L. C. Lew Yan Voon and M. Willatzen, J. Appl. Phys. **109**, 031101 (2011)
[6] J. Pal, G. Tse, V. Haxha, M. A. Migliorato, S. Tomić, Phys. Rev. B **84**, 085211 (2011)
[7] G. Vaschenko, C. S. Menoni, D. Patel, C. N. Tomé, B. Clausen, N. F. Gardner, J.Sun, W. Götz, H. M. Ng and A. Y. Cho, Phys. Stat. Sol. B **235**, 238 (2003)
[8] D. Cai and G-Y. Guo, J. Phys. D: Appl. Phys. **42**,185107 (2009)
[9] K. Shimada, T. Sota, K. Suzuki and H. Okumura, Jpn. J. Appl. Phys. **37**, Pt.2,12A (1998)
[10] G. Vaschenko, D. Patel, C. S. Menoni, N. F. Gardner, J. Sun, W. Go¨tz, C. N. Tome´ and B.Clausen, Phys. Rev. B **64**, 241308 (2001)
[11] O. Ambacher, J. Majewski, C. Miskys, A. Link, M. Hermann, M. Eickhoff, M. Stutzmann, F. Bernardini, V. Fiorentini, V. Tilak, B Schaff and L F Eastman, J.Phys.: Condens. Matter **14**, 3399 (2002)
[12] M. A. Migliorato, D. Powell, A. G. Cullis, T. Hammerschmidt and G. P. Srivastava, Phys. Rev. B **74**, 245332 (2006)
[13] G. Bester, X. Wu, D. Vanderbilt, and A. Zunger, Phys. Rev. Lett. **96**, 187602 (2006)
[14] R. Garg, A. Hüe, V. Haxha, M. A. Migliorato, T. Hammerschmidt and G. P. Srivastava, Appl. Phys. Lett. **95**, 041912 (2009)
[15] W. A. Harrison, Electronic Structure and Properties of Solids, Dover Publications Inc., New York, (1989).
[16] N. Troullier and J. L. Martins, Phys. Rev. B **43**, 1993 (1991).
[17] D. R. Hamann, Phys. Rev. B **40**, 2980 (1989)
[18] S. J. Clark, M. D. Segall, C. J. Pickard, P. J. Hasnip, M. J. Probert, K. Refson, M. C. Payne, Zeitschrift für Kristallographie **220**(5-6) pp.567-570 (2005)
[19] J. P. Perdew and A. Zunger, Phys. Rev. B **23**, 5048 (1981).
[20] H. J. Monkhorst and J. D. Pack, Phys. Rev. B **13**, 5189 (1976)
[21] M.V. Barry, Proc. R. Soc. Lond. A **392**, 45 (1984)
[22] F. Bernardini, V. Fiorentini, D. Vanderbilt, Phys. Rev. B **56**, R10024 (1997)
[23] A. Zoroddu, F . Bernardini, P. Ruggerone and V. Fiorentini, Phys. Rev. B **64**, 045208 (2001)
[24] S. Y. Karpov, Phys. Status Solidi C **7**, 1841 (2010)
[25] S. Q. Wang, H. Q. Ye, J. Phys. Condens. Matter, **17**, 4475 (2005)

3rd Workshop on Theory, Modelling and Computational Methods for Semiconductors IOP Publishing
Journal of Physics: Conference Series **367** (2012) 012007 doi:10.1088/1742-6596/367/1/012007

Strain in inhomogeneous InAs/GaAs quantum dot structures

B. Lassen[1], D. Barettin[2], and M. Willatzen[1]

[1] Mads Clausen Institute for Product Innovation - University of Southern Denmark Alsion 2, DK-6400 Sønderborg, DENMARK

[2] Dept. of Electron. Eng. - Univ. of Rome Tor Vergata Via del Politecnico 1 - 00133 Rome, Italy.

E-mail: benny@mci.sdu.dk

Abstract. Most Non-destructive experimental approaches for the determination of indium concentration profiles give information about average indium concentration profiles only. Due to this, there is a need to extrapolate the indium concentration profiles in a way that takes into account the geometry of the quantum dots. We here present two extrapolation approaches. In the first approach we assume that the indium concentration profile is constant in the direction perpendicular to the measurement plane, while in the second approach we take into account the symmetry of the structure. Both approaches are compared to a profile with a constant indium concentration inside the dot.

1. Introduction

It is well known that InAs/GaAs quantum dot structures grown based on, for example, the Stranski-Krastanow growth mode [1] do not possess a homogeneous indium concentration inside the quantum dot [2]. Specifically, it has been observed that the indium concentration is highest at the tip of pyramidal quantum dots and decreases as we approach the base of the pyramid. Taking into account the shape of the indium concentration profile is important for accurate prediction of electronic properties for two reasons. First, within the k.p approximation [11] the concentration directly affects the effective masses and confinement potentials resulting in shifts in the energies and dipole transition strengths. Secondly, the concentration profile also affects the strain distribution and through that the piezoelectric potential giving additional shifts in energies and changes in transition strengths.

There exists today many approaches for the determination of concentration profiles in quantum dot structures such as cross-sectional transmission electron microscopy (X-TEM) [3], cross-sectional scanning tunneling microscopy (X-STM) [4], scanning transmission electron microscopy (STEM) [5], X-ray photoelectron microscopy [6], anomalous X-ray scattering (AXS) [7], Scanning-Probe-Microscopy nanotomography [8], and composition evaluation by lattice fringe analysis (CELFA) technique [9]. The ultimate goal of all such approach is to provide information about the full three dimensional concentration profile, however, at present the only way of gaining access to this is by destructive means. The non-destructive methods either provide average information in a give cross-section or the concentration profile on the surface of the structure. In this work we focus on the extrapolation of an average concentration profile obtained using the CELFA technique and study the consequences of a given choice of extrapolation.

Journal of Physics: Conference Series **367** (2012) 012007 doi:10.1088/1742-6596/367/1/012007

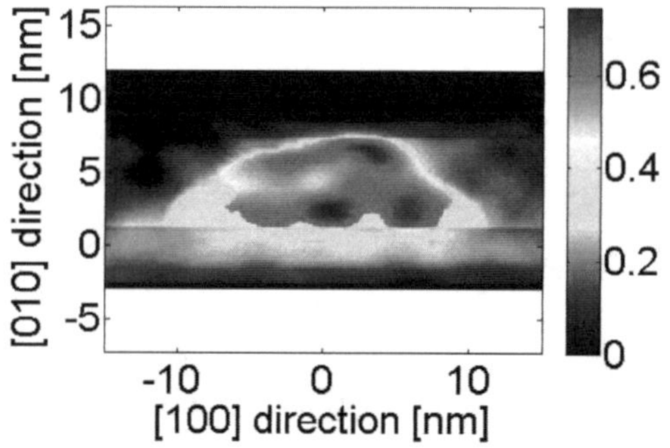

a)

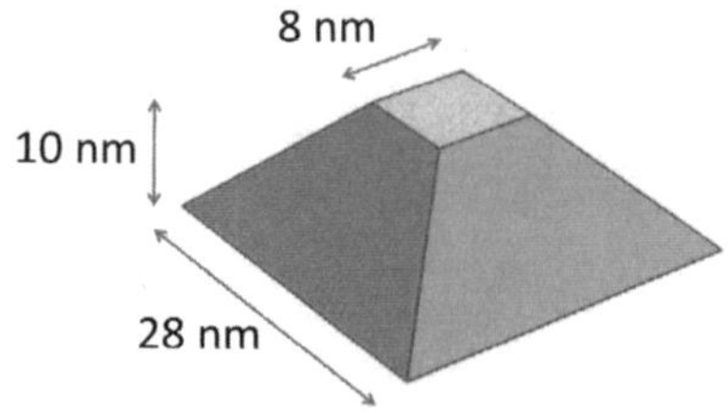

b)

Figure 1. a) Color-coded local In-concentration of a quantum dot as obtained by CELFA (Taken from Blank et al. [9]). b) the geometry of the quantum dot under consideration. The x, y, and z directions are the [100], [010], and [001] directions, respectively.

2. Concentration Profile

In the paper by Blank et al. [9] information about the indium concentration in InAs/GaAs quantum dots has been obtained using the composition evaluation by lattice fringe analysis (CELFA) technique. The problem with the indium concentration profile found using CELFA is that it only gives information about the average indium concentration profile along one direction (in figure 1a the [010] direction).

In this work we consider the following three different indium concentration profiles extracted based on the average indium concentration shown in Figure 1a. All three profiles has been chosen so that their total indium content is identical.

2.1. Profile 1

We assume the indium concentration to be constant in the [010] direction.

$$C(x,y,z) = \begin{cases} \hat{C}(x,z)/L(z) & \text{inside the pyramid} \\ 0 & \text{outside} \end{cases}$$

where $L(z) = (L_d - 2z)$ for $0 < z < L_d/2$ and zero otherwise, $\hat{C}$ is the average indium concentration shown in Figure 1a, and L_d is the side length of the pyramid at $z = 0$. We divide by $L(z)$ in order to take into account the shape of the quantum dot, see reference [9] for further details.

2.2. Profile 2

We assume a symmetric indium concentration profile in the [100] and [010] directions. Given the average concentration profile shown in figure 1 the three dimensional concentration profile takes the form

$$C(x,y,z) = \frac{\hat{C}(x,z)\hat{C}(y,z)}{\int \hat{C}(x,z)dx}$$

where again $\hat{C}$ is the average indium concentration. We divide with $\int \hat{C}(x,z)dx$ in order to ensure that $\int C(x,y,z)dx = \hat{C}(y,z)$.

2.3. Profile 3

We assume a constant indium concentration profile throughout the quantum dot.

$$C(x,y,z) = \begin{cases} C_{tot}/Vol_d & \text{inside the pyramid} \\ 0 & \text{outside} \end{cases}$$

Journal of Physics: Conference Series **367** (2012) 012007 doi:10.1088/1742-6596/367/1/012007

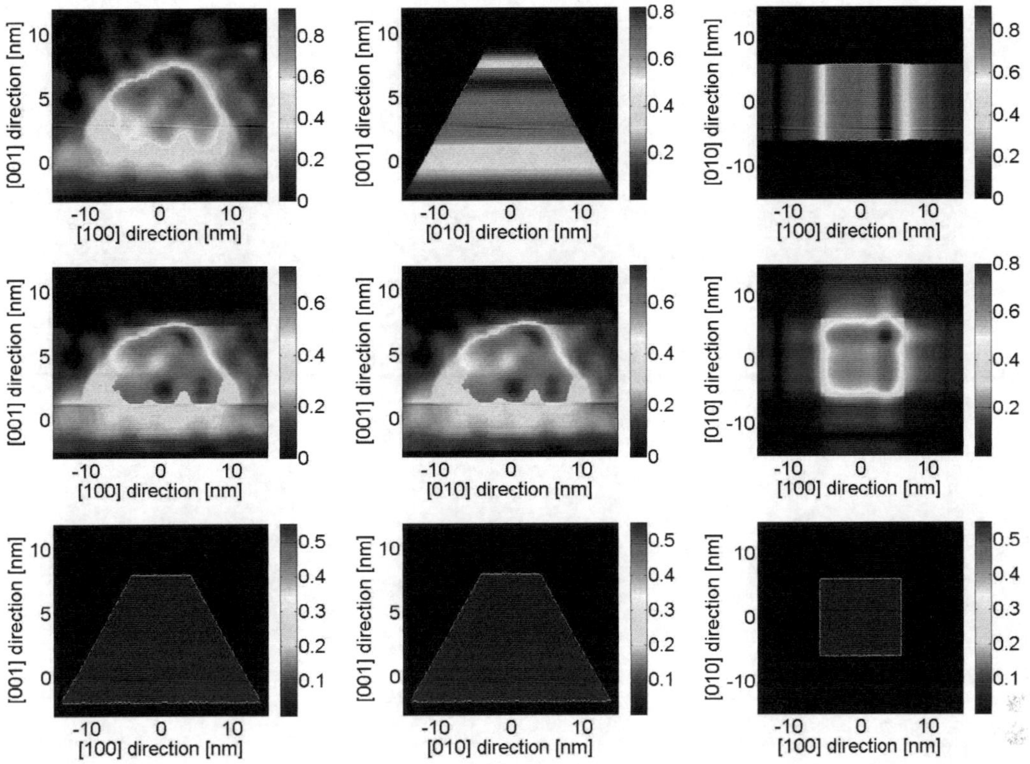

Figure 2. In column 1, 2, and 3 the concentration profile 1, 2 and 3 are shown, respectively. The left plot is the (010) plane, the middle plot the (100) plane, both in the center of the dot, and the right plot is the (001) plane at an elevation of 6 nm.

where $C_{tot} = \int \hat{C}(x,z)dxdz$ and Vol_d is the volume of the pyramid. We have included this profile as this is the most widespread choice of concentration profile for band structure calculations.

In figure 2 we show the three concentration profiles in the (100), (010), and (001) planes. In the figure we see that the first profile (constant along the [010] direction) gives rise to a higher indium concentration at the top of the structure compared with the symmetric profile. It is clear that in the real structure there is no reason why the indium concentration should be constant along the [010] direction.

3. Results

In this work we model strain and piezoelectric effects according to the fully-coupled electromechanical model described in Barettin et al. [10] and the band structure model is the usual Zinc-blende 8 band k·p model (see e.g. [11]) with parameters taken from the paper by Stier et al. [12].

3.1. Electromechanical Effects

In figures 3 we show the hydrostatic strain component $H = \epsilon_{xx} + \epsilon_{yy} + \epsilon_{zz}$ for the three indium concentration profiles. We clearly see the choice of indium concentration profile reflected in the hydrostatic strain. The most prominent differences are seen in the directions where a constant indium concentration is assumed, i.e., when comparing profile 1 with profile 2 the main differences are seen in the [010] direction. Additionally, we see that for profile 1 the strain is highest at the top of the pyramid, whereas for profile 2 we also see high strains further down in the structure, due to high indium concentration further down in the structure for profile 2.

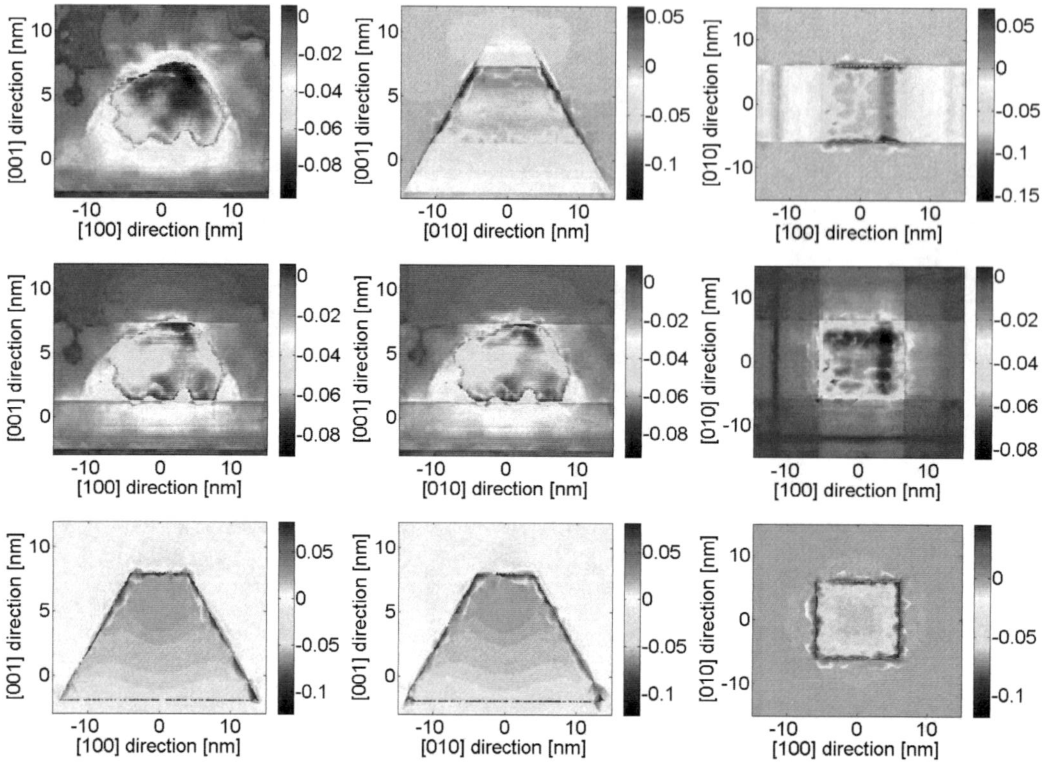

Figure 3. Hydrostatic strain component for indium concentration profile 1. In column 1, 2, and 3 the hydrostatic strain component for profile 1, 2 and 3 are shown, respectively. The left plot is the (010) plane, the middle plot the (100) plane, both in the center of the dot, and the right plot is the (001) plane at an elevation of 6 nm.

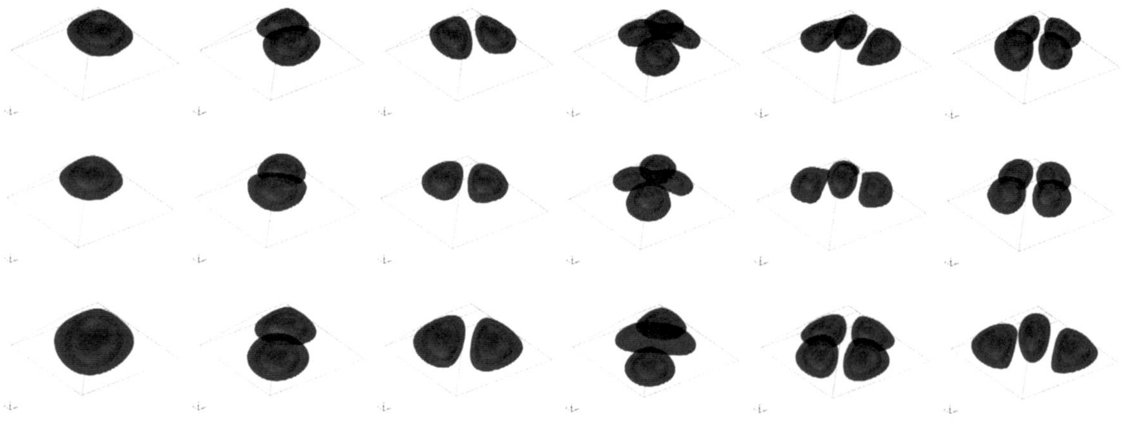

Figure 4. In column 1, 2, and 3 we show the six first conduction band electron states for indium concentration profiles 1, 2, and 3, respectively.

3.2. Electronic Structure

In figure 4 we show the six first conduction band electron states for the three indium concentration profiles. The corresponding energies are given in table 1 below. From the figures we see that the states

Profile 1	0.990	1.040	1.047	1.088	1.105	1.111
Profile 2	1.024	1.074	1.079	1.121	1.134	1.138
Profile 3	1.015	1.038	1.056	1.069	1.093	1.097

Table 1. Conduction band electronic energies for the three profiles. Energies are listed in eV.

for profile 1 and 2 are very similar, the main difference being a small angular rotation of the states. However, the states for profile 3 deviate notable from the states of the two other profile. The states being more spread out and the states 5 and 6 are interchanged compared with the states of the other profiles.

Comparing the conduction band energies of profiles 1 and 2, listed in table 1, we notice that the energy levels of profile 2 is shifted upwards with around 30 meV, however, the spacing of the energy levels of the two profile are more or less the same. Concerning profile 3, we see that here the energy level spacing deviates from the spacings observed for profile 1 and 2. Finally we notice that for profile 2 there are states which are almost degenerate, whereas, none of the states for profile 1 and 3 are close to each other. The reason for this is that concentration profile 2 is much closer to having inversion symmetry than the other two profiles, and as a result we expect to see more states closer to each other (in the case with an inversion symmetric dot the states would be degenerate).

4. Conclusions

We have studied three possible choice of extrapolated concentration profiles based on an average concentration profile obtained using the CELFA technique. The three dimensional concentration profiles were used in determining the strain and piezoelectric fields in the structure and these were included in 8 band k.p calculations. We observed that there were notable differences between the electronic structure of the different profiles. The most notable being shifts in energy levels and lifting of degeneracies due to inversion asymmetric profiles. We furthermore observed a crossing of excited energy levels in going to a constant concentration profile.

References

[1] C. Priester and M. Lannoo, Phys. Rev. Lett. 75, 93 (1995)
[2] D. Litvinov, H. Blank, R. Schneider, D. Gerthsen, T. Vallaitis, J. Leuthold, T. Passow, A. Grau, H. Kalt, C. Klingshirn, and M. Hetterich, J. Appl. Phys. 103, 083532 (2008)
[3] M. Floyd, Y. Zhang, K. P. Driver, J. Drucker, P. A. Crozier, and D. J. Smith, Appl. Phys. Lett., 82, 1473 (2003)
[4] P. Offermans, P. M. Koenraad, J. H. Wolter, K. Pierz, M. Roy, and P. A. Maksym, Phys. Rev. B, 72, 165332 (2005)
[5] A. Rosenauera, T. Mehrtensa, K. Mllera, K. Griesa, M. Schowaltera, P. V. Satyama, S. Bleya, C. Tessareka, D. Hommela, K. Sebalda, M. Seyfrieda, J. Gutowskia, A. Avramescub, K. Englb, S. Lutgenb, Ultramicroscopy, 111, 1316 (2011)
[6] G. Biasiol, S. Heun, G. B. Golinelli, A. Locatelli, T. O. Mentes, F. Z. Guo, C. Hofer, C. Teichert, and L. Sorba, Appl. Phys. Lett., 87, 223106 (2005)
[7] R. Magalhes-Paniago, G. Medeiros-Ribeiro, A. Malachias, S. Kycia, T. I. Kamins, R. Stanley Williams, Phys. Rev. B, 66, 245312 (2002)
[8] A. Rastelli, M. Stoffel, A. Malachias, T. Merdzhanova, G. Katsaros, K. Kern, T. H. Metzger, and O. G. Schmidt, nano lett., 8, 1404 (2008)
[9] H. Blank, D. Litvinov, R. Schneider, D. Gerthsen, T. Passow, and K. Scheerschmidt,Cryst. Res. Technol., 10, 1 (2009)
[10] D. Barettin, S. Madsen, B. Lassen. and M. Willatzen, Commun. Comput. Phys., 11, 797 (2011)
[11] L. C. Lew Yan Voon and M. Willatzen, The k·p method, Springer, 2009
[12] O. Stier, M. Grundmann, and D. Bimberg, Phys. Rev. B, 59, 5688 (1999)

Journal of Physics: Conference Series **367** (2012) 012008 doi:10.1088/1742-6596/367/1/012008

Electron-LO Phonon and Electron-Photon Interactions Analogy in Semiconductor Quantum Dots

Nikola Prodanović

E-mail: elnpr@leeds.ac.uk
Institute of Microwaves and Photonics, School of Electronic and Electrical Engineering, University of Leeds, Woodhouse Lane, Leeds, LS2 9JT, UK

Zoran Ikonić

Institute of Microwaves and Photonics, School of Electronic and Electrical Engineering, University of Leeds, Woodhouse Lane, Leeds, LS2 9JT, UK

Dragan Indjin

Institute of Microwaves and Photonics, School of Electronic and Electrical Engineering, University of Leeds, Woodhouse Lane, Leeds, LS2 9JT, UK

Paul Harrison

Institute of Microwaves and Photonics, School of Electronic and Electrical Engineering, University of Leeds, Woodhouse Lane, Leeds, LS2 9JT, UK

Abstract. We discuss the relationship between the two most important types of intersublevel coupling in the polar semiconductor based quantum dots (QDs), namely Frölich electron-longitudinal optical(LO) phonon coupling and electron-light coupling. We further parametrize the Frölich coupling, giving its quantitative description in terms of QD geometric and composition parameters.

1. Introduction

Electron-longitudinal optical phonon interaction was found to be the most important intrinsic interaction in the conduction band of self-assembled QDs based on polar semiconductors. Either considered as a weak interaction modelled by Fermi golden rule [1] or as a strong interaction leading to the formation of finite-lifetime polarons [2, 3], it is responsible for the decay of excited carriers from p-like first excited state to s-like ground state [4, 5]. Therefore, understanding this interaction and its relations to other relevant quantities is of great importance for understanding the physics of conduction band electrons in QDs. In the following sections we show that this coupling can be parametrized so to relate it to optical matrix elements, giving a clear relationship between these two types of coupling of the electronic degrees of freedom, and providing a

Journal of Physics: Conference Series **367** (2012) 012008 doi:10.1088/1742-6596/367/1/012008

simple intuitive picture of the dependance of Frölich coupling on QD geometric and composition parameters.

2. Initial Considerations

Consider a simple case of one spinless electron in the conduction band of a QD. At low temperatures the electron predominantly occupies s-like ground state. The strongest resonance from that state via optical excitation is with the p-like first excited state [6, 7]. In this reduced space of only two relevant electronic levels an optical LO phonon mode $\boldsymbol{k}$ couples electron wavefunctions via Hamiltonian matrix element $\mathcal{H}^{ph}_{\rm ab}(\boldsymbol{k}) = K\langle\psi_{\rm a}\left|\frac{e^{i\boldsymbol{k}\cdot\boldsymbol{r}}}{k}\right|\psi_{\rm b}\rangle$, where $K = \sqrt{\frac{e^2\hbar\omega_{\rm LO}}{2V}\left(\frac{1}{\epsilon_\infty} - \frac{1}{\epsilon_{st}}\right)}$. This matrix element will be called Frölich coupling function (FCF) for a wavevector $\boldsymbol{k}$. The most important parameter describing this interaction in both weak coupling and strong coupling regime, for such a two-level system, is the integral over the mode space $\boldsymbol{k}$ of the Frölich matrix element [3, 8], $C^2_{\rm ab} = \int {\rm d}^3\boldsymbol{k}\left|\mathcal{H}_{\rm ab}(\boldsymbol{k})\right|^2$, which will be denoted as Frölich coupling constant.

The electron-photon interaction is described within the dipole approximation. The optical matrix element coupling these two states in the dipole approximation is $\boldsymbol{\mathcal{R}}_{\rm ab} = \langle\psi_{\rm a}\left|\boldsymbol{r}\right|\psi_{\rm b}\rangle$.

3. The Relationship

Due to confinement, only LO-phonons with small wavevectors can significantly interact with electrons. By expanding the exponential term in FCF in Taylor series around $\boldsymbol{k} = 0$ one obtains

$$\begin{aligned}
\mathcal{H}^{\boldsymbol{k}}_{\rm ab} &= K\int {\rm d}^3\boldsymbol{r}\psi^{(\rm a)*}\frac{\cos\boldsymbol{k}\cdot\boldsymbol{r}}{k}\psi^{(\rm b)} + iK\int {\rm d}^3\boldsymbol{r}\psi^{(\rm a)*}\frac{\sin\boldsymbol{k}\cdot\boldsymbol{r}}{k}\psi^{(\rm b)} = \\
&= \frac{K}{k}\int {\rm d}^3\boldsymbol{r}\psi^{(\rm a)*}\psi^{(\rm b)} + iK\boldsymbol{e_k}\cdot\int {\rm d}^3\boldsymbol{r}\psi^{(\rm a)*}\boldsymbol{r}\psi^{(\rm b)} - \\
&- \frac{k}{2}K\int {\rm d}^3\boldsymbol{r}\psi^{(\rm a)*}\left(\boldsymbol{e_k}\cdot\boldsymbol{r}\right)^2\psi^{(\rm b)} - i\frac{k^2}{6}K\int {\rm d}^3\boldsymbol{r}\psi^{(\rm a)*}\left(\boldsymbol{e_k}\cdot\boldsymbol{r}\right)^3\psi^{(\rm b)} + \cdots,
\end{aligned} \tag{1}$$

where the second equation represents the Taylor series of sine and cosine functions, and $\boldsymbol{e_k} = \frac{\boldsymbol{k}}{k}$. In the case of well defined and opposite parity for $|\psi_a\rangle$ and $|\psi_b\rangle$ we have

$$\mathcal{H}^{\boldsymbol{k}}_{\rm ab} = iK\left(\boldsymbol{e_k}\cdot\boldsymbol{\mathcal{R}}_{\rm ab} - \frac{k^2}{6}\int {\rm d}^3\boldsymbol{r}\psi^{(\rm a)*}\left(\boldsymbol{e_k}\cdot\boldsymbol{r}\right)^3\psi^{(\rm b)} + \cdots\right). \tag{2}$$

In the limit $k \to 0$ the expression (2) becomes:

$$\mathcal{H}^{\boldsymbol{k}}_{\rm ab} = iK\boldsymbol{e_k}\cdot\boldsymbol{\mathcal{R}}_{\rm ab}. \tag{3}$$

Hence, the maximum of the scalar field $\mathcal{H}^{\boldsymbol{k}}_{\rm ab}$ is proportional to the intensity of the dipole coupling vector, and occurs at $k \to 0$.

FCF is not well defined at $k = 0$ where the vector $\boldsymbol{k}$ does not have a defined direction, and therefore the expression $\boldsymbol{e_k}\cdot\boldsymbol{\mathcal{R}}_{\rm ab}$ is not defined. Strongly anisotropic behaviour of the distribution $\mathcal{H}^{\boldsymbol{k}}_{\rm ab}$ arises from the factor $\boldsymbol{e_k}\cdot\boldsymbol{\mathcal{R}}_{\rm ab}$. For $\boldsymbol{k}$ pointing in the direction of the dipole coupling vector, the FCF exhibits the weakest negative slope. This slope is increasing with increasing angle between $\boldsymbol{k}$ and the FCF, and decays rapidly to zero when $\boldsymbol{k}$ becomes almost perpendicular to the dipole coupling vector, and in the limit where $\boldsymbol{k}$ is exactly perpendicular the FCF equals zero. All these features prove that the FCF has a p-orbital like shape. This was indeed expected, since the FCF is essentially the Fourier transform of an odd function.

Therefore, the Frölich coupling function has been parameterized by a bell-shaped distribution function, with the maximal value at $k \to 0$ proportional to the intensity of the dipole coupling vector and multiplied by the $\frac{k_i}{k}$, where i denotes the direction of the dipole coupling vector. This model gives:

$$\left|\mathcal{H}_{\mathrm{ab}}^{\boldsymbol{k}}\right| = |\boldsymbol{\mathcal{R}}_{\mathrm{ab}}|\, K \frac{k_z}{\sqrt{k_x^2+k_y^2+k_z^2}} \mathcal{I}(\boldsymbol{k}), \tag{4}$$

where $\mathcal{I}(\boldsymbol{k})$ is the anisotropic distribution function with maximum $\mathcal{I}(0) = 1$.

The general shape of the distribution function $\mathcal{I}(\boldsymbol{k})$ reduced to two dimensions with HWHMs σ_x and σ_z is shown in Fig. 1. The maximal value is always $\mathcal{I}(\boldsymbol{k}=0) = 1$. The exact lineshape

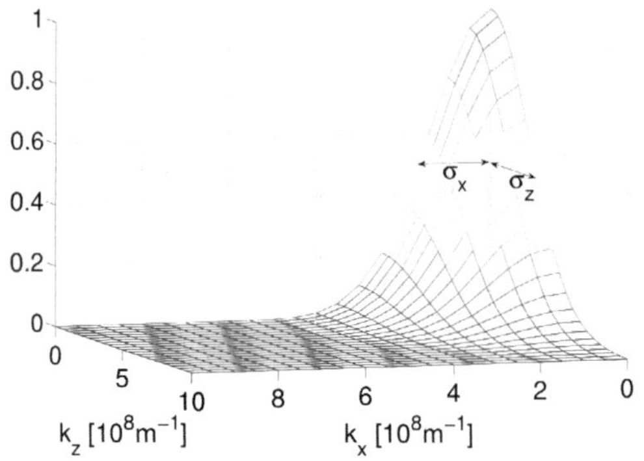

Figure 1. The distribution function for a parabolic QD. The figure is reduced to two dimensions (z and x) without losing generality. Linewidths σ_x and σ_z are inversely proportional to dot sizes in the corresponding directions.

depends on the particular QD and its geometric and composition parameters. A more extended dot in a particular direction has a narrower distribution function in that direction, i.e. σ_x and σ_z are decreasing functions of the dot sizes in x and z directions. This is clear from the fact that FCF is a Fourier transform of the product of the s-like and p-like wavefunction divided by k. It is well known that the transform of an expanded function shrinks, and vice versa. For the parabolic potential QD the distribution function is a Gaussian with HWHMs inversely proportional to the extensions of the parabolic confinement. By changing the QD material composition, i.e. increasing the dot depth, one inreseas the confinement of the bound wavefunctions. Therefore, for the same reason as above, the HWHM is an increasing function of the dot depth.

The FCF has been parametrized via two factors, namely the dipole coupling vector $\boldsymbol{\mathcal{R}}_{\mathrm{ab}}$, which is a well known spectroscopic quantity, and the distribution function with the property $\mathcal{I}(0) = 1$. Using this approximaton Frölich coupling constant becomes:

$$C_{\mathrm{ab}}^2 = \frac{e^2\hbar\omega_{\mathrm{LO}}}{16\pi^3}\left(\frac{1}{\epsilon_\infty} - \frac{1}{\epsilon_{st}}\right)|\boldsymbol{\mathcal{R}}_{\mathrm{ab}}|^2 \int \mathrm{d}^3\boldsymbol{k} \frac{k_z^2}{k_x^2+k_y^2+k_z^2}\mathcal{I}^2(\boldsymbol{k}) \tag{5}$$

In calculating the Frölich coupling constant, the exact lineshape of the distribution function loses its significance due to the integration. The HWHM of the distribution function and the dipole coupling vector pre-factor remain as the main factors affecting the value of Frölich coupling constant. The integral in the last equation is affected directly by variations of QD geometric and composition parameters. Being the Fourier-transform-like image of the product of s-like and p-like wavefunctions, it shrinks and expands by expansion and reduction of the QD size. The dipole coupling vector can, at the same time, exhibit different behaviour, preventing the entire FCF to change significantly. However, this will still influence the QD optical properties. The larger the magnitude of the dipole coupling vector, the stronger is the s-p-like optical resonance.

4. Conclusion

In summary, we have parameterized the FCF via the dipole coupling vector (optical matrix element) and the distribution function whose widths were related to geometric and composition properties of the QD. There are a number of advantages of such parameterization. By using the expansion formula for $\mathcal{H}_{\rm ab}(\boldsymbol{k})$ one can significantly reduce the numerical effort in calculation of all coupling constants relevant to electron-phonon interaction. The presented model can be used as the basis for estimating the ratio between radiative and non-raditive lifetimes as a function of the dot geometric and composition parameters, which is very important in future designs of QD based intersublevel emitters. By measuring the absorption one can extract the values of the dipole coupling vector and the level separation, which can be further used in combination with the presented model to get an insight into the electron-LO phonon interaction.

References

[1] Bockelman U and Bastard G 1990 *Phys. Rev. B* **42** 8947–8951
[2] Hameau S, Guldner Y, Verzelen O, Ferreira R, Bastard G, Zeman J, Lemaître A and Gérard J M 1999 *Phys. Rev. Lett.* **83** 4152–4155
[3] Stauber T, Zimmermann R and Castella H 2000 *Physical Review B* **62**(11) 7336–7343
[4] Zibik E A, Wilson L R, Green R P, Bastard G, Ferreira R, Phillips P J, Carder D A, Wells J P R, Cockburn J W, Skolnick M S, Steer M J and Hopkinson M 2004 *Phys. Rev. B* **70** 161305
[5] Grange T, Ferreira R and Bastard G 2007 *Phys. Rev. B* **76**(24) 241304
[6] Ulbrich N, Bauer J, Scarpa G, Boy R, Schuh D, Abstreiter G, Schmult S and Wegscheider W 2003 *Applied Physics Letters* **83** 1530–1532
[7] Fischer C, Bhattacharya P and Yu P C 2003 *Electronics Letters* **39** 1537 – 1538
[8] Zibik E A, Grange T, Carpenter B A, Porter N E, Ferreira R, Bastard G, Stehr D, Winnerl S, Helm M, Liu H Y, Skolnick M S and Wilson L R 2009 *Nature Materials* **8** 803–807

Journal of Physics: Conference Series **367** (2012) 012009 doi:10.1088/1742-6596/367/1/012009

Photon Absorption in Periodically Regimented Nanostructures

A Luque-Rodríguez, F M Gómez-Campos, S Rodríguez-Bolívar
Departamento de Electrónica y Tecnología de los Computadores,
Facultad de Ciencias, Av. Fuentenueva s/n, CP 18071, Universidad de Granada, Spain

E-mail: fmgomez@ugr.es

Abstract. Calculations based on the Effective Mass Approach of the miniband structure in the conduction band, joint density of states and photon absorption coefficient for InAs quantum dots embedded into a GaAs matrix are presented. In our investigation we kept the same volume for the quantum dots (250 nm^3) and the separation between quantum dots (3 nm) and changed the aspect ratio in order to determine its effects on the quantities. From the results we draw conclusions that could enlighten the physical processes about photon absorption in regimented nanostructures.

1. Introduction

From the first proposal of theoretical one-dimensional superlattices by L. Esaki and R. Tsu [1] to nowadays, electronics have witnessed revolutionary changes in experimental and theoretical fields, and many researchers in the world have contributed to perform what once was considered just a speculative model. In the recent years two- and three-dimensional regimented arrays of nanostructures has attracted great attention by experimental [2-3] and theoretical [4-7] groups. Therefore, prediction on the physics of such systems is needed to gain insight on how they would work.

In this work we present a study of the confinement and photon absorption by regimented arrays of quantum dots. We calculated the miniband structure, joint density of states (JDOS) and photon absorption for arrays of cuboid InAs quantum dots embedded into a GaAs matrix. For the sake of clarity we did this study for quantum dots having the same volume, but changing their shapes.

2. Theoretical Formalism

The theory for two or three-dimensional periodical arrays of quantum dots share common features with that of regular crystals. Schrödinger equation for these systems is:

$$-\frac{\hbar^2}{2}\left[\nabla\frac{1}{m(\vec{r})}\nabla\right]\Psi_{\vec{q},n}(\vec{r})+V_{strained}(\vec{r})\Psi_{\vec{q},n}(\vec{r})=E_n(\vec{q})\Psi_{\vec{q},n}(\vec{r}) \tag{1}$$

where $m(\vec{r})$ is the space-dependent effective mass, $V_{strained}(\vec{r})$ is the band profile (from a single band approach) where strain could be considered, $\Psi_{\vec{q},n}(\vec{r})$ is the carrier wave function and $E_n(\vec{q})$ the corresponding eigenvalue. According to Bloch-Floquet theorem, the above $\Psi_{\vec{q},n}(\vec{r})$ has the following structure

$$\Psi_{\vec{q},n}(\vec{r})=\eta_{\vec{q},n}(\vec{r})e^{i\vec{q}\cdot\vec{r}} \qquad (2)$$

where $\eta_{\vec{q},n}(\vec{r})$ is a function having the periodicity of the superlattice and $e^{i\vec{q}\cdot\vec{r}}$ is a plane wave with a phase related to a $\vec{q}$ vector of the superlattice reciprocal space, the so-called Q space. For a given $\vec{q}$, the $\eta_{\vec{q},n}(\vec{r})$ function was expanded using 2197 plane waves (13 in each direction) within the normalized plane wave approach and the resulting Hamiltonian was diagonalized to obtain both $E_n(\vec{q})$ and $\Psi_{\vec{q},n}(\vec{r})$. The absorption coefficient $\alpha(\hbar\omega_{op})$ was evaluated using the method presented elsewhere [4] about photon absorption in periodic nanostructures.

We have done the study for quantum dots having three different shapes: flattened cuboids (L_x=10 nm, L_y=10 nm, L_z=2.5 nm), regular cubes (L_z=6.3 nm, L_y=6.3 nm, L_z=6.3 nm), and elongated cuboids (L_x=5 nm, L_y=5 nm, L_z=10 nm). All of them have the same volume (250 nm^3) and the separation between neighbour quantum dots is 3 nm for every case. The dots were aligned along the X, Y and Z directions in a tetragonal crystal system. We used the notation for the relevant points in the Q space following the usual name given in the literature for tetragonal crystal systems [8].

3. Results

In this section we show the main results obtained from this study about miniband structure, JDOS and photon absorption.

3.1. Miniband structure and JDOS

Figure 1 shows the miniband structure belonging to the conduction band for the three samples. The lowest value of the first miniband is practically the same for the three cases. The width of the first miniband, i.e. the spreading in energy from the lowest to the highest value within a miniband, goes from 152 meV for the flattened cuboid down to 103 meV for the regular cube case. The miniband structure is crucial to understand the photon absorption process. It is observed how the quantum dot side lengths play a relevant role in the miniband profiles and the degeneracy of the bands.

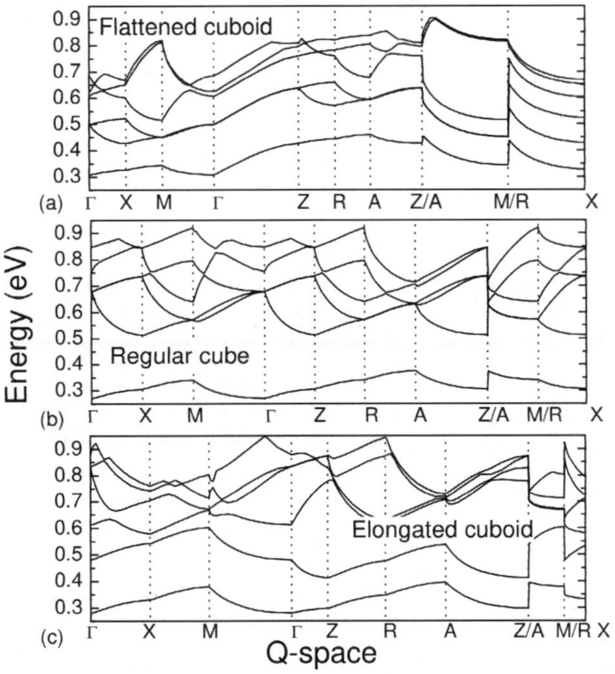

Figure 1. Miniband structures for quantum dot arrays with three dots having different shapes (sizes in the text).

It is clearly shown that a gap between first and second miniband appears. Selection rules for photon absorption imply $\vec{q}$ vector conservation [4] and therefore the transitions for optical absorption are vertical in Figure 1. Similar gaps are found between higher minibands (for instance, between third and fourth minibands for the flattened cuboid), but the occupancy of these minibands are low and their contribution is negligible in the absorption process. The distance between the first and second miniband could give us an approximated idea about the absorption energies. Regular cube shows the higher vertical separation between first and second miniband (~0.25eV) while the lowest separations are for the flattened cuboid (~0.1eV).

In order to understand the whole photon absorption process, a rough approximation to approach this phenomenon is to consider transitions just between the first and second miniband, assuming a fully occupied first miniband and empty second miniband. Although this is a rough modelling of the problem, this would help us to study these structures in the first steps, and afterwards we shall show more accurate calculations. With this aim we calculated the JDOS for transitions between the first and second minibands and show it in Figure 2. One can identify an absorption threshold for elongated and flattened cuboids about 100 meV and about 200 meV for regular cubes. In the next section we show these values are relevant to understand the absorption coefficient.

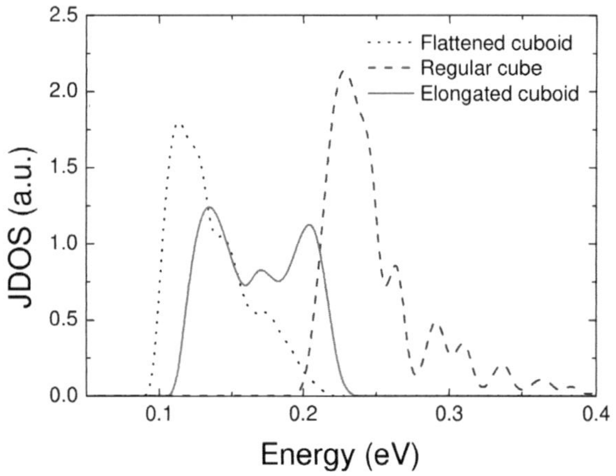

Figure 2. Joint density of States (JDOS) computed for transitions between the first and second minibands for the three studied cases.

3.2. Absorption coefficient.

Finally, the absorption coefficient is computed taking into account the ten lowest energy minibands, a light polarized in (1,1,1) direction, and the occupancy factor for each miniband. The latter is a function of the Fermi level. In this work we placed it at the InAs midgap. However, since the calculations where done for a Fermi level far from the carrier states, a shift of ΔE in the position of the Fermi level just implies a change in the value of the absorption coefficient (~$\exp(\Delta E/kT)$) [9], thus the absorption coefficient profile is almost unaffected. In other words, the absorption could be more or less intense but the threshold and transmission energies would be the same. In Figure 3 the absorption coefficient for the three cases is depicted, where a threshold value is found for the absorption coefficient in agreement with those observed in the JDOS. This means that a simple model considering only the two first minibands and computing the JDOS helps to obtain the threshold energies.

Regarding to the absorption coefficient magnitude, a sample with a lower first miniband implies more proximity to the Fermi level, and hence higher occupied states and higher number of available transitions. This is the case of the regular cube and elongated cuboid (Figures 1b and 1c) which have their first miniband slightly lower than the flattened cuboid (Figure 1a), slightly reducing the absorption coefficient values.

It is worth to note the double peak that appears in the absorption coefficient for the elongated cuboid. The peak at low energies is due to transitions between the first and second miniband (Figure

1c). The second peak is for transitions between the first miniband and the third and higher minibands. The gap between the second and third minibands (Fig. 1c) corresponds in the absorption coefficient with the deep valley between the two peaks.

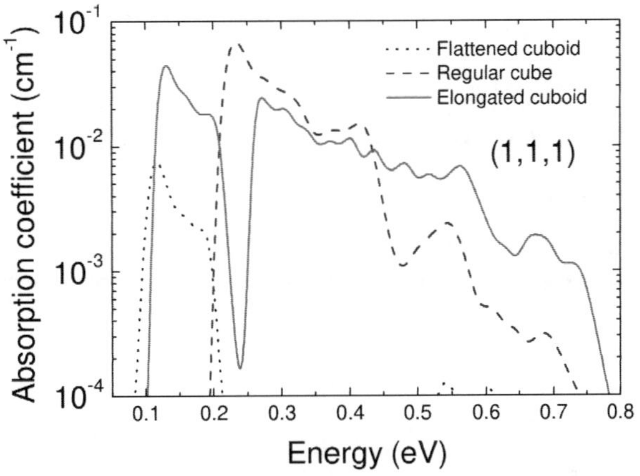

Figure 3. Comparison between the absorption coefficients obtained for the three different shapes for an incident light polarized in (1,1,1).

4. Conclusions

The calculation of the miniband structure and photon absorption coefficient in regimented structures of InAs quantum dots in a matrix of GaAs has been presented. The quantum dot volume has been kept constant at 250 nm^3 and three different shapes have been analyzed. The most important feature is the existence of a threshold in the absorption coefficient about photon energies lying in the far infrared region of the spectrum. The thresholds are closely related to the miniband structure and JDOS for transitions between the first and second miniband. According to these predicted features, these periodic systems might be of application in far infrared photon detectors.

Acknowledgments

This work was supported by Ministerio de Educación y Ciencia and Fondo Europeo de Desarrollo Regional (FEDER) under research projects TEC2010-16211.

[1] Esaki L and Tsu R 1970 *IBM J. Res. Dev.* **14** 61
[2] Verma V B, Dias N L, Reddy U, Bassett K P, Li X, Coleman J J 2011 *IEEE J. Quantum Elect.* **47** 417
[3] Murphy J E, Beard M C, Norman A G, Ahrenkiel S P, Johnson J C, Yu P, Mićić O I, Ellingson R J, Nozik A J 2006 *J. Am. Chem. Soc.* **128** 3241
[4] Gómez-Campos F M, Rodríguez-Bolívar S, Luque-Rodríguez A, López-Villanueva J A, Carceller J E 2010 *J. Appl. Phys.* **108** 124307
[5] Tomic S, Jones T, Harrison N 2008 *Appl. Phys. Lett.* **93** 263105.
[6] Kłos J W, Krawczyk M 2010 *J. Appl. Phys.* **107** 043706
[7] Gómez-Campos F M, Rodríguez-Bolívar S, de Jong van Coevorden C M, Luque-Rodríguez A, Lara-Bullejos P, Carceller J E 2009 *Phys. E: Low-dim. Syst. Nanostruct.* **41** 1712
[8] Setyawan W, Curtarolo S 2010 *Comp. Mater. Sci.* **49** 299
[9] Rodríguez-Bolívar S, Gómez-Campos F M, Luque-Rodríguez A, López-Villanueva J A, Jiménez-Tejada J A, Carceller J E 2011 *J. Appl. Phys.* **109** 074303

Journal of Physics: Conference Series **367** (2012) 012010 doi:10.1088/1742-6596/367/1/012010

Accurate light scattering for non spherical particles from Mie-type theory

B. Hourahine, K. Holms and F. Papoff
Department of Physics, University of Strathclyde, 107 Rottenrow, Glasgow G4 0NG, UK
E-mail: benjamin.hourahine@strath.ac.uk
E-mail: f.papoff@strath.ac.uk

Abstract. We report a new approach for accurate calculation of optical cross sections and internal and scattered fields at any point in space for micro- and nanoparticles. Our approach is based on constructing the intrinsic optical modes of general smooth particles and hence optimised surface Green functions that, for any incident field, provide an *a priori* upper bound on the error and identify the class of incident fields with largest error.

Non-spherical particles with at least one characteristic length on the order of the wavelength of incident light have remarkable optical properties: metallic particles can enhance the surface field by several orders of magnitude [1]; dielectric particles can have exceptionally high quality factors [2] or focus light into sub-wavelength photonic jets [3]. This underpins a large amount of applied and fundamental research, ranging from various types of surface enhanced nonlinear spectroscopy [4] to single photon light-matter interaction [5]. The optical response of these particles depends both on their shape and constituent material, with internal and scattered electromagnetic fields being solutions of the macroscopic Maxwell equations that satisfy the appropriate boundary conditions at the surface of the particle and at infinity. The great variety of particle shapes used in experiments and the lack of an exact theory analogous to Mie's solution for spheres leads to the use of fully numerical or semi-analytical methods in theoretical investigations [6, 7]. In order to systematically study the dependence of near and far field optical properties on shape and input field, we need a method that fulfils several requirements. To increase the accuracy of the calculated fields the error with which the boundary conditions are fulfilled must be minimised. Furthermore, as this error depends on the incident field, we need an *a priori* estimate of the error at any point in space for *any* incident field if predictions are to be considered reliable. Finally, it is desirable that calculation of the resulting fields arising for different input fields must have a low computational cost for their evaluation at any point in space. In this paper we describe an approximate theory in which, like Mie theory, all fields are expanded as sums of vector functions, and each function in the incident field is associated with only one function in both the internal and scattered fields. This approach fulfils all the requirements mentioned above. In the rest of the paper we outline our method, derive an exact formula for the *a priori* error for any incident field, and compare this method with selected particles where it is possible to obtain essentially exact results. We conclude with an outline of future developments and with a brief comparison of our method with the most common alternatives.

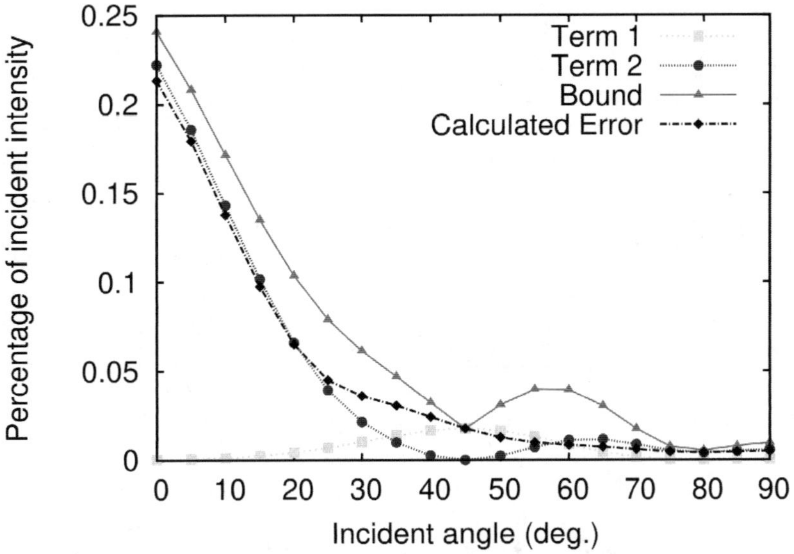

Figure 1. Surface error for a rod shaped metallic particle of diameter one wavelength (700 nm) and aspect ratio two, illuminated by 45° polarised incident plane waves. The GSGF was optimised for an incident field at 45°; the error bound and the actual evaluated error match at this angle, where the error is due to the part of the field inside the optimising set but outside the range of the Green's function. (term 1 on the right-hand of Eqn. 2). At 0° incidence the error is clearly seen to be due term 2 of Eqn. 2, implying that this field should also be included in the set over which the GSGF is optimised. For incident angles larger than 45°, the error is of the same order as the 45° incident field within $\{\mathcal{O}^0\}$.

In the following we use the Einstein convention of summation over repeated indexes, unless otherwise specified, and the compact notation $\mathcal{F}^\sigma(\mathbf{x}) = [\mathbf{E}^\sigma(\mathbf{x}), \mathbf{H}^\sigma(\mathbf{x})]^T$, with $\sigma = 0, i, s$ for incident, internal and scattered electromagnetic fields respectively. We indicate the field components of $\mathcal{F}^\sigma$ tangent to the particle surface S at point s as $\mathbf{f}^\sigma(s)$. We use $\langle \mathbf{a}|\mathbf{b}\rangle = \int_S \mathbf{a}^* \cdot \mathbf{b}\, ds$ and $\|\mathbf{a}\| = \langle \mathbf{a}|\mathbf{a}\rangle^{1/2}$ for scalar products and norms over S. Dielectrics without losses and metals are modelled by real and complex dielectric constants respectively [8]. For these media, the boundary conditions are that the energy of the scattered field flows toward infinity and the components of the total external and internal fields tangent to the surface of the particle are identical. Electromagnetic fields at the surface of the particle and within the internal and external regions are given by the surface integral $\mathcal{F}^\sigma(\mathbf{x}) = \int_S G^\sigma(\mathbf{x}, s)\mathcal{F}^0(s)ds$, where G is the dyadic surface Green function [9, 10] and $\sigma = i(s)$ labels the internal (scattered) field.

We have previously [11] approximated G with an operator that we call a Generalised Surface Green Function (GSGF), because they are based on generalised inverses, using these GSGFs to calculate internal and scattered field of elongated particles for incident fields propagating along the symmetry axis. We now substantially generalise that approach by optimising the GSGF over a set of incident fields, $\{\mathcal{O}^0\}$, and provide a rigorous *a priori* upper bound on the error everywhere in space and for any incident field. We approximate internal and scattered fields by finite combinations of regular and radiating electric and magnetic multipolar fields, respectively [12, 13]. These are exact solutions of the Maxwell equations and are linearly independent and complete over the surface of the particle [8]. Note that the surface residual $\|\mathbf{f}^0 - \mathbf{f}^i + \mathbf{f}^s\|$ (for Lyapunov [14] surfaces or those continuous up to at least the second derivative) provides an upper bound to the maximum error on scattered and internal fields that decreases with the distance from S (see Eqn. 3.3 of Ref. [8]), as long as the error in the propagation, i.e., in the evaluation of the multipolar functions used to expand of solutions of the Maxwell equations, is negligible. For this reason the coefficients c_j of the fields' expansion that minimise the residual provide an optimal description of the fields at any point. These coefficients are the solutions of the self-adjoint linear system $\langle \mathbf{f}_k|\mathbf{f}_j\rangle c_j = \langle \mathbf{f}_k|\mathbf{f}^0\rangle$, but solving this system directly leads to large numerical errors [15]. We address this problem by sampling the tangent components of the of electric and magnetic fields at a set of points on the surface, $\{\mathbf{\Pi}\}$, and then constructing orthonormal sets of internal and scattering tangential fields, $\{\mathbf{i}\}$ and $\{\mathbf{s}\}$, which span the same spaces as the functions $\{\mathbf{f}^i\}$ and $\{\mathbf{f}^s\}$ at those points, via singular value decomposition [16] of the sampled functions and subsequent unitary transformation to pairs of *principal* vectors [17, 18]. The resulting pairs of internal and scattered functions are orthogonal to all but at most one function in the other space, and oriented at an angle ξ to their partner ($\mathbf{i}_j \cdot \mathbf{s}_j = \cos(\xi_j)$). We can then find the internal and scattered fields for a particular incident field by decomposing it into a sum of pairs of principal modes, $\mathbf{f}^0 = \sum_j a^i_j \mathbf{i}_j - a^s_j \mathbf{s}_j$, where $a^{i/s}$ are amplitudes of each of the internal and scattered principal modes for that specific incident field [18]. This approach provides a substantial improvement in numerical stability compared to the original self-adjoint linear system, combined with a determination of the *intrinsic* principal modes of the scatterer.

A further important difference with respect to previous work [11, 20] is that the surface points are chosen such that Gaussian quadrature [21] can be applied, resulting in errors in scalar products and norms on S due to discretisation that can be safely neglected as they are of an order lower ($o(\epsilon)$) than the leading error, ϵ, of this theory, which itself is due to the use of finite numbers of functions to expand the fields. By changing the number of functions used in the expansions of $\mathbf{f}^{i,s}$, how close the zeros and singularities of the expanding functions are to S and the number of principal mode pairs (r) included in the solution, we minimise the surface error over a set of different incident fields $\{\mathcal{O}^0\}$, so that we have $\|\mathbf{o}^0_j - \mathbf{f}^i + \mathbf{f}^s\| \leqslant \epsilon \|\mathbf{o}^0_j\|$ for each input field $\mathcal{O}^0_j$. Here ϵ is one or two orders of magnitude smaller than the threshold below which the scattering differential cross section converges [20].

The GSGF can also be constructed from the principal modes of the particle; for a set of modes optimised over $\{\mathcal{O}^0\}$ with leading error ϵ, the GSGF is

$$G^\sigma(\mathbf{x}, s) = \sum_{j \leq r} \frac{1}{\sin^2 \xi_j} \begin{cases} +\mathcal{I}_j(\mathbf{x})\langle i_j - \cos(\xi_j) s_j| & ; \sigma = i \\ -\mathcal{S}_j(\mathbf{x})\langle s_j - \cos(\xi_j) i_j| & ; \sigma = s \end{cases} . \tag{1}$$

Note that Mie theory can be recast in the form of Eqn. 1 with vector spherical harmonics and spherical vector wavefunctions for incident and scattered light respectively [18].

The advantage of using a set of exact solutions of the Maxwell equations to expand the principal modes is twofold: the resulting fields can be propagated analytically anywhere in space at virtually no computational cost and some field features can be found analytically. Furthermore, from Eqn. 1 we can derive an exact bound for the surface residual for any incident field, including those outside of $\{\mathcal{O}^0\}$ in terms projectors into and outside of the space of $\{\mathcal{O}^0\}$, $P_{\{\mathcal{O}^0\}} = |\mathbf{o}_i\rangle\langle\mathbf{o}_i|\mathbf{o}_j\rangle^{-1}\langle\mathbf{o}_j|$ and $P_{\perp\{\mathcal{O}^0\}} = \mathbf{1} - P_{\{\mathcal{O}^0\}}$:

$$\|\mathbf{f}^0 - \mathbf{f}^i + \mathbf{f}^s\| \leqslant \|(\mathbf{1} - G^S_{\{\mathcal{O}^0\},\epsilon}) P_{\{\mathcal{O}^0\}}|f^0\rangle\| + \|(\mathbf{1} - G^S_{\{\mathcal{O}^0\},\epsilon}) P_{\perp\{\mathcal{O}^0\}}|f^0\rangle\| , \tag{2}$$

where $G^S = -\hat{\mathbf{n}}\times\hat{\mathbf{n}}\times[G^i(s', s) - G^s(s', s)]$, with G^S providing the response of the particle through projection of incident fields onto the sets $\{\mathbf{i}\}$ and $\{\mathbf{s}\}$, while $(\mathbf{1} - G^S_{\{\mathcal{O}^0\},\epsilon})$ is the orthogonal projector into the kernel of $G^S_{\{\mathcal{O}^0\},\epsilon}$. The first term on the right of Eqn. 2 gives the error for the part of an incident field within $\{\mathcal{O}^0\}$ which is lost under the action of the Green's function, while the second term is the contribution from the part of the incident field outside of the

a)

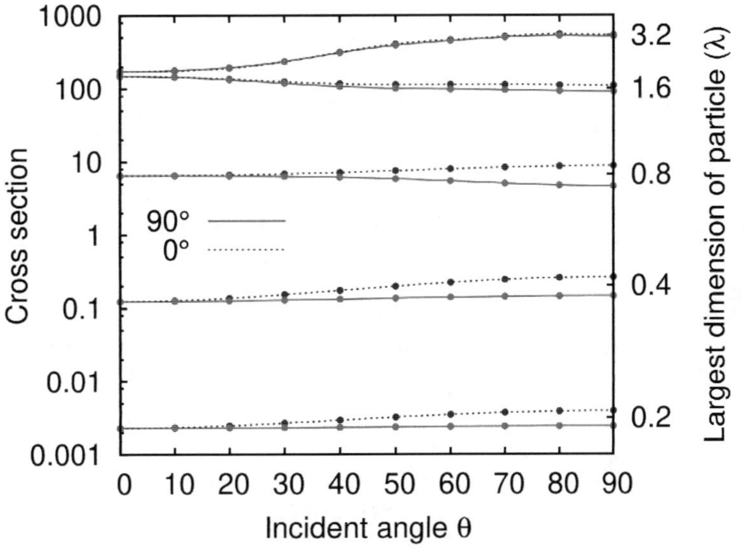

b)

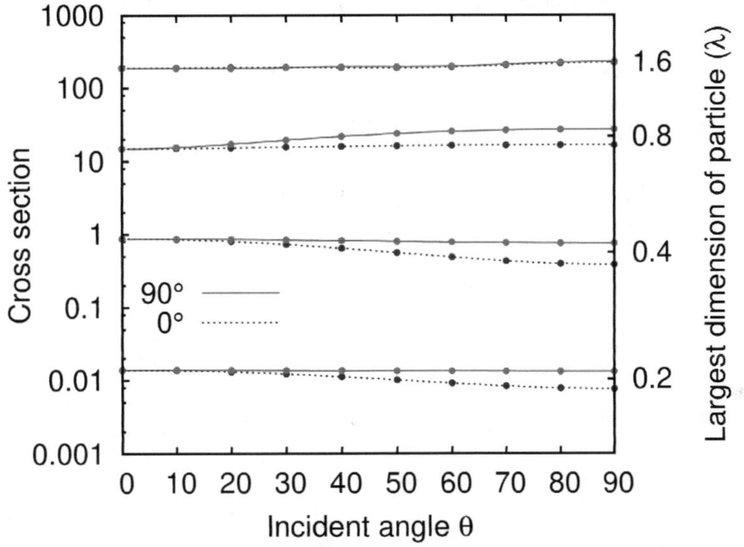

Figure 2. Scattering cross sections for a) Prolate and b) oblate spheroids with aspect ratios of 2 and a refractive index of 1.5. Both planes of light polarisation with respect to the axis of symmetry are used (0° and 90°). Results are shown for the current method (points) and a Mie-like separation of variables method (lines) [19].

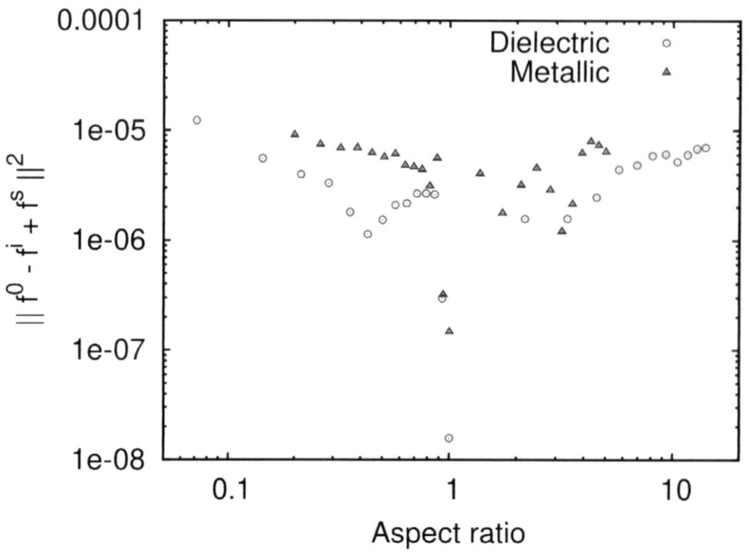

Figure 3. Variation in optimised surface error for families of dielectric (open circles, refractive index $n = 1.5$) and gold (triangles, refractive index $n \sim 0.32 + 3.18i$) rounded disc and rod-like particles as a function of the ratio between the semi-diameters parallel and perpendicular to the symmetry axis. Aspect ratio 1 corresponds to a sphere with radius $r = 1/2\pi$ wavelengths.

optimisation set which is also outside of the space of the principal modes. The surface residual for a field for which the GSGF has not been optimised is then of the same order ϵ as for a field in $\{\mathcal{O}^0\}$ when the second term of Eqn. 2 is $O(\epsilon)$. In other words, we have identified the part of a generic input field that may reduce the precision of the scattering calculation: it is the input field component that is orthogonal to the set $\{\mathcal{O}^0\}$ and within the kernel of the GSGF. This is important not only theoretically, but also practically, even without optimisation. When the GSGF is not optimised, the set $\{\mathcal{O}^0\}$ corresponds to the finite set of incident fields in which the error is measured directly, but one can estimate the error for an arbitrary incident field once the error has been tested on a finite set of incident fields. In Fig. 1 we show the two right-hand terms of Eqn. 2 and their sum, as compared against the calculated error for an example particle where the Green's function was optimised for a single incident field incident at 45° with respect to the particle axis.

Once the error is low, the solutions are highly reliable. Figure 2 demonstrates the agreement between the results of our theory, and the essentially exact results from the generalisation of Mie theory to one of the few other cases where coordinate separation is possible [19]. Finally in Fig. 3 we demonstrate the range of particles that can be reliably treated with our approach, showing the maximum value of ϵ^2 for a range of rounded dielectric and metallic rods and disks; minimising the residual over a set of plane waves linearly polarised at 45° to the plane of incidence, with angles of incidence to the symmetry axis ranging from 0° to 90° at intervals of 15°. Note that the limiting case of rods or disks with aspect ratio 1 is a sphere with radius $1/2\pi$ in units of the wavelength of the incident field.

In conclusion, our formalism, as in Mie theory, identifies the contribution to internal and scattered field of each of the functions in the expansion of the incident field and allows us to study the dependence of optical properties of resonant particles on their shape and on the input field. For the determination of internal and scattered fields of symmetric particles over large regions of space, the computational complexity of our method is similar to that of the T-matrix and Discrete Sources [20] approaches, but the principal modes have the advantage of allowing a great compression of information and reduction of memory cost. This approach is less demanding than methods such as the Finite Domain Time Difference (FDTD), because the fields are propagated everywhere simply by evaluating the special functions contained in the multipolar fields. The advantage of our approach with respect to the Discrete Dipole Approximation (DDA) is that the number of multipolar fields necessary for the GSGF is smaller than the number of dipoles used in the DDA. Furthermore, GSGF does not suffer from stair-casing errors and field approximations at metal/dielectric interfaces, as the FDTD and DDA, and have the unique features of providing equations and diagnostic for the error and of allowing a partial analytical analysis. The derivation of GSGF for layered particles, particles with inclusions and particles clusters is straightforward, while GSGF for metallic particles with low symmetry and sharp edges are much more challenging; in these cases the FDTD and DDA [6] remain the methods of choice.

References

[1] R. Hillenbrand and F. Keilmann. Optical oscillation modes of plasmon particles observed in direct space by phase-contrast near-field microscopy. *Appl. Phys. B*, 73(3):239–243, September 2001.
[2] K. J. Vahala. Optical microcavities. *Nature*, 424:839, 2003.
[3] S. Kong, S. A. Sahakian, A. Taflove, and V. Backman. Photonic nanojet-enabled optical data storage. *Optics Express*, 16(18):13713–13719, 2008.
[4] D. Graham and R. Goodacre. Chemical and bioanalytical applications of surface enhanced Raman scattering spectroscopy. *Chem. Soc. Rev.*, 37:883–884, 2008.
[5] Takao Aoki, Barak Dayan, E. Wilcut, W. P. Bowen, A. S. Parkins, T. J. Kippenberg, K. J. Vahala, and H. J. Kimble. Observation of strong coupling between one atom and a monolithic microresonator. *Nature*, 443:671–674, 2006.
[6] M.I.Mishchenko, J. H. Hovernier, and L. D. Travis, editors. *Light scattering by nonspherical particles: Theory, Measurements and Applications.* Academic Press, 2000.
[7] M. Karamehmedović, R. Schuh, V. Schmidt, T. Wriedt, C. Matyssek, W. Hergert, A. Stalmashonak, G. Seifert, and O. Stranik. Comparison of numerical methods in near-field computation for metallic nanoparticles. *Opt. Expr.*, 19:8939–8953, 2011.
[8] A. Doicu, T. Wriedt, and Y. Eremin. *Light Scattering by Systems of Particles.* Springer, 2006.
[9] T. Rother, M. Kahnert, A. Doicu, and J. Wauer. Surface Green's Function of the Helmholtz Equation in Spherical Coordinates. *Prog. Electromag. Res.*, 38:47–95, 2002.
[10] T. Rother. Self-consistent green's function formalism for acoustic and light scattering, part 1: Scalar notation. *Optics Communications*, 251:254–269, 2005.
[11] K. Holms, B. Hourahine, and F. Papoff. Calculation of internal and scattered fields of axisymmetric nanoparticles at any point in space. *J. Opt. A: Pure Appl. Opt.*, 11:054009, May 2009.
[12] A. Doicu and T. Wriedt. Calculation of the t matrix in the null-field method with discrete sources. *J. Opt. Soc. Am. A*, 16(10):2539–2544, OCT 1999.
[13] A. Doicu and T. Wriedt. Extended boundary condition method with multipole sources located in the complex plane. *Optics Communications*, 139:85–91, 1997.
[14] E.D. Solomentsev. Encyclopedia of mathematics, http://www.encyclopediaofmath.org/.
[15] J. A. Eremin, N. V. Orlov, and V. I. Rozenberg. Scattering by nonspherical particles. *Comp. Phys. Commun.*, 79:201–214, 1994.
[16] J. Gentle. *Numerical Linear Algebra for Applications in Statistics.* Springer-Verlag, Berlin, 1998.
[17] C. Jordan. Essai sur la géométrie à n dimension. *Buletin de la Société Mathématique de France*, 3:103–174, 1875.
[18] F. Papoff and B. Hourahine. Geometrical mie theory for resonances in nanoparticles of any shape. *Opt. Expr.*, 19:21432–21444, 2011.
[19] N. V. Voshchinnikov and V. G. Farafonov. Optical properties of spheroidal particles. *Astrophysics and Space Science*, 204(1):19–86, 1993.
[20] E. Eremina, Y. Eremin, and T. Wriedt. Analysis of the light scattering properties of a gold nanorod on a plane surface via discrete sources method. *Optics Communications*, 267:524–529, 2006.
[21] W. H. Press, B. P. Flannery, S. A. Teukolsky, and W. T. Vetterling. *Numerical Recipes in FORTRAN.* Cambridge University Press, Cambridge, England, 2 edition, 1992.

Journal of Physics: Conference Series **367** (2012) 012011 doi:10.1088/1742-6596/367/1/012011

Intersubband Thermophotovoltaics

M.F. Pereira[1,2]

[1]Materials and Engineering Research Institute, Sheffield Hallam University
S1 1WB, Sheffield, United Kingdom
[2]Department of Physics, Jazan University, Jazan, Saudi Arabia

mpereira@shu.ac.uk

Abstract. In this paper valence band designs that allow simple normal incidence in the TE mode as in a convention solar cell are studied and shown to be more efficient than conduction band-based TM structures in a range of the far infrared. The relevance of many-body corrections is further investigated numerically for III-V quantum well structures.

1. Introduction

Increasing the efficiency of photovoltaic (PV) materials and devices is one of the most important current challenges for optical sciences and technology [1]. One of the possible ways to improve the efficiency is by absorbing photons in the mid and far infrared for which conventional semiconductors are transparent. Different solutions have been proposed, e.g. up/down conversion layers [2] or introducing intermediate bands [3]. All of those have their own efficiency limitations and technological difficulties. One other solution, discussed in this paper is the concept of intersubband (ISB) thermophotovoltaic (TPV) devices, in which the transition energies leading to absorption of radiation do not depend on the band gap [4]. The present state-of-the-art in TPV [5] energy conversion from a thermal source at a temperature of 1300 K is about 0.8 W/cm2, obtained with a 0.6-eV bandgap energy InGaAs interband photodetector grown lattice mismatched on InP with an InPAs buffer and the preliminary simulations of Ref. [6] indicate that at the same temperature the ISB TPV can generate 1.4W/cm2. We can therefore conclude that the multiple-junction ISB device concept has the potential to provide a substantial improvement over existing technologies.

In this paper we analyze a different approach: valence-band based designs that lead to TE mode absorption and can directly absorb photons normal to the surface in contrast to usual conduction band based ISB devices that operate with TM incidence leading to additional complications like the requirement of prism and/or couplers. Furthermore, many body effects due to the electron-electron interaction that have a proven relevance in semiconductor lasers in both interband [7-14] and intersubband architectures [15-25]. Their relevance for TPV structures is made clear here and this may be particularly important for hot carrier architectures, where a large density of nonequilibrium carriers can be created [26].

Journal of Physics: Conference Series **367** (2012) 012011 doi:10.1088/1742-6596/367/1/012011

2. Numerical Method and Applications

The photocurrent generation spectrum [27] collected by each period (n) with length d_n, which here is the quantum well (QW) width, is given by

$$J_n(\omega) = q \int_{z_n}^{z_n+d_n} G_n(\omega, z)dz, \tag{1.a}$$

where the generation rate is

$$G_n(\omega, z) = \alpha_n(\omega)\Upsilon(\omega)[1 - R(\omega)]exp(-\alpha_n d_n), \tag{1.b}$$

In Eq. (1), $R(\omega)$is the surface reflection coefficient, $\Upsilon(\omega)$ is the photon flux incident on the sample surface, and z_n is the position of the QW along the device's growth direction.

The absorption $\alpha(\omega)$, or gain spectra $g(\omega) = -\alpha(\omega)$ are calculated by means of a Keldysh nonequilibrium Green's functions approach in which the susceptibility $\chi(\omega)$ is directly related to the carriers Green's functions G, which satisfies a Dyson equation.

$$\alpha(\omega) = \frac{4\pi\omega}{cn_b} \mathrm{Im}\{\chi(\omega)\},\ \chi(\omega) = \frac{2}{V}\sum_{\mu\neq\nu,\vec{k}} \wp_{\mu\nu}(k)\chi_{\mu\nu}(k). \tag{3}$$

where n_b denotes the background refractive index, c is the speed of light, and $\wp_{\mu\nu}$is the transition dipole moment between subbands ν and μ. The nonequilibrium steady-state susceptibility function in Eq. (3) stems from the carrier Keldysh Green's function. The numerical scheme used here can be briefly summarized as follows. The first step is the solution of the 8x8 k·p Hamiltonian [4, 17, 25, 28] describing the heterostructure without the Coulomb interaction. The Green's functions and self-energies are expanded using eigenstates and eigenvalues of this Hamiltonian. Next, by assuming thermalized electrons or holes, the full NEGF scheme is simplified and reduces to the self-consistent evaluation of chemical potentials and self-energy matrix elements which lead to subband energy renormalizations, dephasing constants, and occupation functions. Only carrier-carrier scattering is considered here and details of the corresponding self-energy are given in Ref. 14. Finally, absorption and gain are given by the solution of the integrodifferential equation obtained from the carriers Green's function by numerical matrix inversion [17,18].

The numerical results presented next are for a single junction section composed of a 5 nm GaAs – Al0.3Ga0.7As quantum well. In all curves the valence band (for TE polarization) and conduction band (for TM polarization) electrons available due to doping are initially thermalized at 300K.

Figure 1 compares and contrasts the Photocurrent Generation Spectra with and without many body corrections at the same level as in Ref. [17].

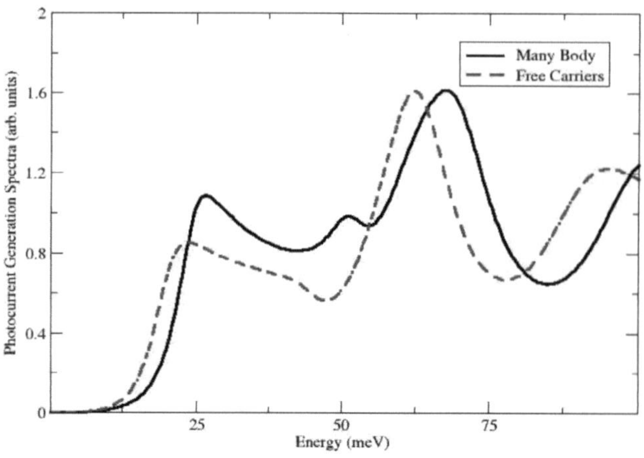

Figure 1. TE Photocurrent generation spectra with (solid) and without manybody effects (dashed) for one section of a possible ISB device with a 5 nm quantum well active region. The electrons in the quantum well valence bands are thermalized at T=300K and the source is at T=750K. The carrier density is N $=3\times10^{12}$ /cm^2.

Figure 2 compares and contrasts the photocurrent for conduction band based (TM mode) and valence band based structures (TE mode). Even though in the mid infrared the much larger TM dipole moment more than compensates the losses due to the required prisms and/or couplers to bring TM polarized radiation to the active region, further in the infrared the TE mode dominates. This may be very useful to extract energy from thermal sources that emit strongly in that range.

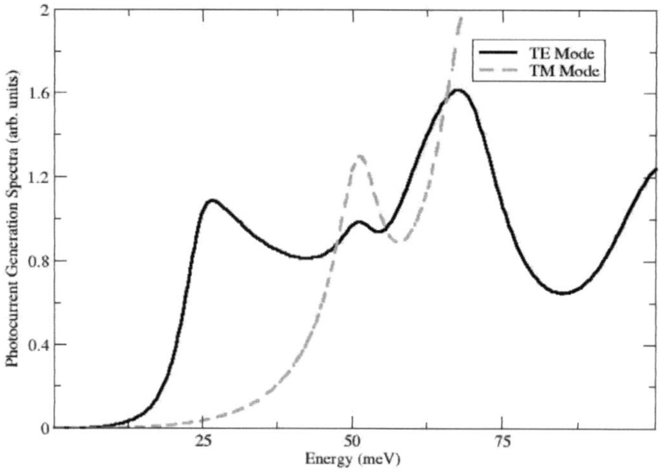

Figure 2. Photocurrent generation spectra for the QW of Fig.1 under same conditions. The solid lines are for TE polarization and the dashed for TM at a maximum angle of incidence. The actual absorption would be smaller since a non-zero angle is required for TM absorption. Thus the contrast would be even better in practice. Note that for TM polarization we assume doping in the conduction bands and for TE in the valence bands.

At this point a few remarks are due related to the choice of materials and structures. The transition dipole moment is rigorously obtained through a commutator from the momentum matrix element. A general discussion has shown that for intervalence band transitions three terms contribute [28]. For example, in normal TE incidence (E, =O) and k=0, the last term couples light- and heavy-hole components of the initial and final states, with a strength proportional to the mixed derivatives (xz or yz) in the Hamiltonian. The second term is important only if the initial and final hole states have a significant mix of conduction-band states, which is possible in the 8×8 model as the light-hole bands are coupled to the conduction and spin-orbit bands even at k=0. For nonzero k, the first term allows transitions between same parity components of the initial and final states. For more details see the detailed discussion in Ref. [28]. This means that the second term leads to increased dipole moment for small bandgap materials, which might be an advantage in designing ISB TPVs based on TE mode operation. Larger well widths lead to larger dipoles even in the TM mode, dominated by interconduction band transitions that are not much sensitive to band coupling. In the TE mode a larger well width will further contribute by reducing the confined level separation between conduction and valence band states and thus increasing the dipole moment through bandcoupling in the same way that a reduced bandgap does. However, the Coulomb enhancement is reduced for larger well widths, so for a given target wavelength operation range a detailed analysis of the interplay between bandstructure and many body effects as a function of well width should be made.

In summary, this study demonstrates that inter-valence band absorption structures may lead to devices that can potentially convert thermal photons to current efficiently in the far infrared, which is a range that remains to be further exploited. Absorption due to valence band transitions allows simple perpendicular incidence in the TE mode without the need of prisms or other couplers that increase the complexity, cost, practicality and also the actual losses in a conduction-band-based TPV. Furthermore, the numerical analysis demonstrates the relevance of many body effects, which are now routinely considered in laser and optoelectronic device simulators, but have been essentially overlooked for photo and thermophotovoltaic simulation and design consideration.

References

[1] A. Feltrin and A. Freundlich, Renewable Energy 33, 180-185 (2008).
[2] T. Trupke et al, J. Appl. Phys. 92, 1668 (2007).
[3] S. Tomić, Phys. Rev. B 82, 195321 (2010).
[4] M.F. Pereira Jr., J. Opt. Soc. Am. B 28, 2014 (2011).
[5] B. Wernsman et al," IEEE Trans. Electron. Dev. 51(3), 512 (2004).
[6] J. Yin and R. Paiella , Optics Express 18, 1618 (2010).
[7] W.W. Chow, M.F. Pereira Jr and S.W.Koch. Appl. Phys. Lett. 61, 758 (1992).
[8] M.F. Pereira Jr, R. Binder and S.W. Koch, Appl. Phys. Lett. 64, 279 (1994).
[9] M.F. Pereira Jr and K. Henneberger, phys. stat. sol. 202, 751 (1997).
[10] W.W. Chow, M.F. Pereira Jr and S.W.Koch. Appl. Phys. Lett. 61, 758 (1992).
[11] M.F. Pereira Jr, R. Binder and S.W. Koch, Appl. Phys. Lett. 64, 279 (1994).
[12] M.F. Pereira Jr and K. Henneberger, phys. stat. sol. 202, 751 (1997).
[13] M.F. Pereira Jr and K. Henneberger, phys. stat. sol. (b) 206, 477 (1998).
[14] M.F. Pereira Jr and K. Henneberger, phys. stat. sol. (b) 206, 477 (1998).
[15] G. Scalari et al, Phys. Rev. Lett. 93, 237403 (2004).
[16] J. Li and C.Z. Ning, Phys. Rev. B 70, 125309 (2004).
[17] M.F. Pereira Jr and H. Wenzel, Phys. Rev. B70, 205331 (2004).
[18] M.F. Pereira Jr., S.-C. Lee and A. Wacker, Phys. Rev. B69, 205310 (2004).
[19] R. Nelander et al, Journal of Applied Physics 102, 113104 (2007).
[20] M.F. Pereira Jr. et al, J. Mater Sci: Mater Electron 18, 689 (2007).
[21] M.F. Pereira Jr., Phys Rev B 75, 195301 (2007).

[22] M.F. Pereira Jr., Phys. Rev. B 78, 245305 (2008).
[23] T. Schmielau and M.F.Pereira Jr, Appl. Phys. Lett. 95, 231111 (2009).
[24] Thi Uyen-Khanh Dang et al, Phys. Rev. B 82, 045305 (2010).
[25] M. F. Pereira Jr. and S. Tomić, Appl. Phys. Lett. 98, 061101 (2011).
[26] A. Le Bris and J.F. Guillemoles, Appl. Phys. Lett. 97, 113506 (2010).
[27] V. Aroutiounian, et al. J. Appl. Phys. 89, 2268–2271 (2001).
[28] F. Szmulowicz, Phys. Rev. B51, 1613 (1995).

Compliant energy and momentum conservation in NEGF simulation of electron-phonon scattering in semiconductor nano-wire transistors

J R Barker[1], A Martinez[2] and M Aldegunde[2]

[1]School of Engineering, University of Glasgow, G12 9YH;
[2]College of Engineering, Swansea University, UK.

E-mail: John.Barker@glasgow.ac.uk

Abstract. The modelling of spatially inhomogeneous silicon nanowire field-effect transistors has benefited from powerful simulation tools built around the Keldysh formulation of non-equilibrium Green function (NEGF) theory. The methodology is highly efficient for situations where the self-energies are diagonal (local) in space coordinates. It has thus been common practice to adopt diagonality (locality) approximations. We demonstrate here that the scattering kernel that controls the self-energies for electron-phonon interactions is generally non-local on the scale of at least a few lattice spacings (and thus within the spatial scale of features in extreme nano-transistors) and for polar optical phonon-electron interactions may be very much longer. It is shown that the diagonality approximation strongly under-estimates the scattering rates for scattering on polar optical phonons. This is an unexpected problem in silicon devices but occurs due to strong polar SO phonon-electron interactions extending into a narrow silicon channel surrounded by high kappa dielectric in wrap-round gate devices. Since dissipative inelastic scattering is already a serious problem for highly confined devices it is concluded that new algorithms need to be forthcoming to provide appropriate and efficient NEGF tools.

1. Introduction

The Keldysh non-equilibrum Green function (NEGF) methodology has been applied successfully to the simulation of silicon nano-wire transistors [1-4]. Its success arises from powerful recursive algorithms coupled with the non-perturbative treatment of impurities and surface/interface roughness [1-4] and the easy incorporation of dissipation through diagonal self-energies [4, 5]. The double space representation is ideal for inhomogeneous systems. For steady state models, it is necessary to solve for the energy-resolved two-point Green function $G(\mathbf{x}_1,\mathbf{x}_2;\varepsilon)$ in conjunction with the corresponding self energies. The NEGF formalism is reduced to solving the following system of equations:

$$G^{<}(\mathbf{r}_1,\mathbf{r}_2;\varepsilon) = \iint d^3\mathbf{r}_3 d^3\mathbf{r}_4 G^R(\mathbf{r}_1,\mathbf{r}_3;\varepsilon)\Sigma^{<}(\mathbf{r}_3,\mathbf{r}_4;\varepsilon)G^A(\mathbf{r}_4,\mathbf{r}_2;\varepsilon) \quad (1)$$

$$(\varepsilon\mathbf{1} - \mathbf{H}_0 - \mathbf{V} - \mathbf{e}\Phi - \Sigma^{\mathbf{R}})\mathbf{G}^R(\varepsilon) = \mathbf{1} \quad (2)$$

Inelastic modelling of inelastic scattering in short channel silicon nano-transistors was thought to be irrelevant to the simulation of in-channel processes due to quasi-ballistic coherent flows. In practice,

Journal of Physics: Conference Series **367** (2012) 012012 doi:10.1088/1742-6596/367/1/012012

the presence of strong size quantization/confinement, particularly in wrap-round gate structures (fig 1), can enhance energy dissipation within the various sub-band channels[1, 3].

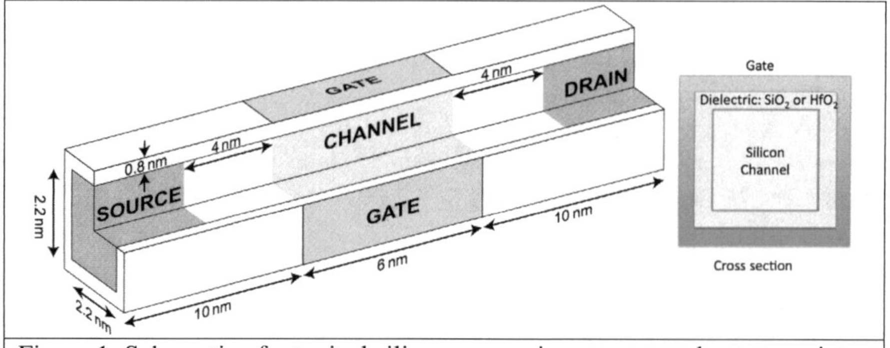

Figure 1: Schematic of a typical silicon nano-wire wrap-round gate transistor

The Green functions then strictly require the inversion of a strongly *non-diagonal* matrix due to the non-diagonality of the self-energies. For silicon it has been argued that a diagonal approximation is sufficient [5]; however, strict energy and momentum conservation is suspect and has led to poor confidence in estimates of channel mobility. The importance of non-diagonal self energies has been pointed out in studies of quantum cascade lasers [6-7] mainly for polar electron-phonon coupling in III-V devices. Recently, the fabrication of wrap-round gate nano-wire transistors (figure 1) with high-κ dielectric surrounding the silicon core leads to consideration of strong SO phonon polar interactions extending into the channel. Here we examine the diagonality issue for some generic examples.

2. Self-Energy for Electron-Phonon interactions

Let us consider the lesser self-energy in the self-consistent Born approximation (SCBA):

$$\Sigma^<(\mathbf{r}_1,\mathbf{r}_2;\varepsilon) = (2\pi)^{-3}\int d^3q\,|U(\mathbf{q})|^2 \exp[i\mathbf{q}.(\mathbf{r}_1-\mathbf{r}_2)] \\ \{n_q G^<(\mathbf{r}_1,\mathbf{r}_2;\varepsilon-\hbar\omega_q)+(n_q+1)G^<(\mathbf{r}_1,\mathbf{r}_2;\varepsilon+\hbar\omega_q)\} \tag{3}$$

To expose the non-diagonality we consider a quasi-1D approach in which the z-direction is the direction of the device spatial inhomogeneity and the x-y plane is homogeneous; we Fourier transform the lateral coordinates to momentum space:

$$\Sigma^<(z_1,z_2;\mathbf{k}_\perp;\varepsilon) = (2\pi)^{-3}\int_{-\pi/a}^{\pi/a} dq_z \int d^2q_\perp\,|U(\mathbf{q})|^2 \exp[iq_z.(z_1-z_2)] \\ \{n_q G^<(z_1,z_2;\mathbf{k}_\perp-\mathbf{q}_\perp;\varepsilon-\hbar\omega_q)+(n_q+1)G^<(z_1,z_2;\mathbf{k}_\perp-\mathbf{q}_\perp;\varepsilon+\hbar\omega_q)\} \tag{4}$$

The non-locality of the self-energy is controlled by the scattering kernel K that we define as:

$$K(z_1,z_2;\mathbf{k}_\perp;\varepsilon) = (2\pi)^{-3}\int_{-\pi/a}^{\pi/a} dq_z\,|U(\mathbf{q})|^2 \exp[iq_z.(z_1-z_2)] \tag{5}$$

Note the phonon wave-vector integration is over the first Brillouin zone and a is the lattice spacing.

3. The Scattering Kernel

The scattering kernel gives a measure of the spatial spread or range of the scattering interaction. For inelastic screened polar optical phonon scattering we find:

$$K = C\int_{-\pi/a}^{\pi/a} dq_z\,|U(\mathbf{q})|^2 \exp[iq_z(z_1-z_2)] \tag{6}$$

$$C = e^2\frac{\hbar\omega_0}{2\varepsilon_0}\left(\frac{1}{\varepsilon_\infty}-\frac{1}{\varepsilon_s}\right); \qquad |U(\mathbf{q})|^2 = C\frac{q_\perp^2+q_z^2}{(q_\perp^2+q_z^2+\xi^{-2})^2} \tag{7}$$

Figure 2 illustrates the scattering kernel plotted as a function of the range in units of the lattice spacing, $R=(z_1-z_2)/a$ for various approximations. First, we replace $|U(\mathbf{q})|^2$ by a constant C_1. This fact this is a very good approximation for inelastic scattering on non-polar optical phonons. The result is a characteristic *sinc* function with zeros at $R=\pm1, \pm2,\ldots$ as shown in Figure 2(a).

Journal of Physics: Conference Series **367** (2012) 012012 doi:10.1088/1742-6596/367/1/012012

$$K = C_1\int_{-\pi/a}^{\pi/a} dq_z \exp[iq_z(z_1 - z_2)] = 2aC_1\frac{\sin[\pi R]}{R} \tag{8}$$

in the approximation that $a \to 0$ the integration extends to infinity, K becomes diagonal :

$$K(R) \to \delta(R) \quad \Rightarrow \Sigma^<(z_1,z_2;\mathbf{k}_\perp,\varepsilon) \sim \delta(z_1 - z_2) \tag{9}$$

Second, we exhibit the full scattering kernel for weakly screened ($\xi/a = 8.9$) polar optical phonon scattering: figure 2(b) gives the exact result using integration over the first Brillouin zone; figure 2(c) shows the approximation $a \to 0$ extending the range of integration; the result closely matches the exact result. Third, the corresponding results for strong screening (shorter screening length $\xi/a = 1.8$) are shown in Figures 2(d) (exact), 2(e) ($a \to 0$ approximation: a considerably poorer approximation here). Clearly, the effective non-locality (non-diagonality) of K and hence Σ extends over a range of the order of the screening length. Similar considerations apply to acoustic phonon scattering. Although it is often assumed that the self-energy is diagonal for high temperature acoustic phonon scattering that is only true for the elastic scattering assumption coupled with an extension of integration beyond the first Brillouin zone. If the scattering treated as inelastic we have for high temperatures:

$$\begin{aligned}\Sigma^<(z_1,z_2;\mathbf{k}_\perp;\varepsilon) &= C_5\int_{-\pi/a}^{\pi/a} dq_z \int d^2q_\perp \exp[iq_z.(z_1 - z_2)] \\ &\{G^<(z_1,z_2;\mathbf{k}_\perp - \mathbf{q}_\perp;\varepsilon - \hbar\omega_q) + G^<(z_1,z_2;\mathbf{k}_\perp - \mathbf{q}_\perp;\varepsilon + \hbar\omega_q)\}\end{aligned} \tag{10}$$

For $\hbar\omega_q = \hbar sq << \varepsilon$ we may pick out the q-dependence as:

$$\begin{aligned}&G^<(z_1,z_2;\mathbf{k}_\perp - \mathbf{q}_\perp;\varepsilon - \hbar\omega_q) + G^<(z_1,z_2;\mathbf{k}_\perp - \mathbf{q}_\perp;\varepsilon + \hbar\omega_q) \approx \\ &2G^<(z_1,z_2;\mathbf{k}_\perp - \mathbf{q}_\perp;\varepsilon) + (\hbar sq)^2\frac{\partial^2 G^<(z_1,z_2;\mathbf{k}_\perp - \mathbf{q}_\perp;\varepsilon)}{\partial\varepsilon^2}\end{aligned} \tag{11}$$

The leading term is the elastic approximation; the second term gives rise to a non-diagonal kernel

$$K_1(|z_1 - z_2|/a) = a^3\int_{-\pi/a}^{\pi/a} dq_z q_z^2 \exp[iq_z(z_1 - z_2)] = \frac{2(\pi R^2 - 2)\sin(\pi R) + 4\pi R\cos[\pi R]}{R^3} \tag{12}$$

This kernel is plotted in figure 3 where it is compared with the elastic approximation (scaled for comparison). Although many diagonality approximations are used for acoustic phonon scattering including the inelastic case, it is evident the approximation fails at short distances.

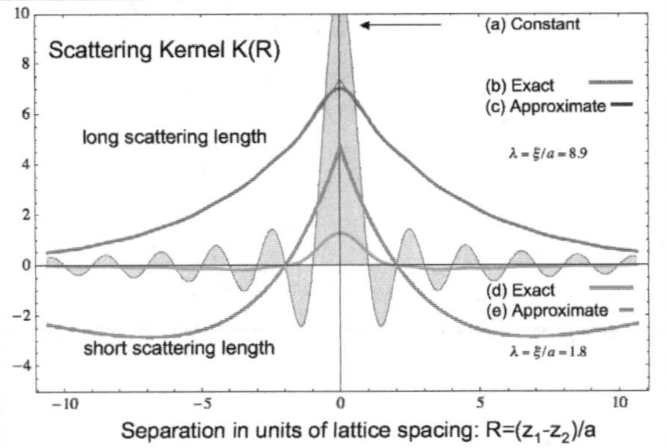

Figure 2: Exact and approximate Scattering Kernel for inelastic scattering on optical phonons

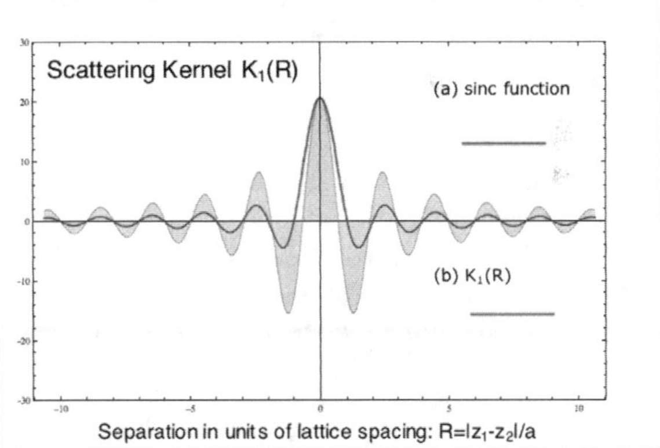

Figure 3: Non-diagonality of the scattering kernel For acoustic phonon scattering

4. Scattering rates and non-diagonality of the self-energy

As a measure of the effects of diagonality/non-diagonality we evaluate the scattering out rate $\Gamma(k_z,k_\perp,\ \varepsilon = \varepsilon[k_z] + \varepsilon[k_\perp])$ for polar optical phonon scattering using (4) and (5) in the self-consistent Born approximation for the retarded self-energy Σ^R (using iteration on the Born approximation result) in the approximation that level shift terms and $G^<$ dependence are neglected:

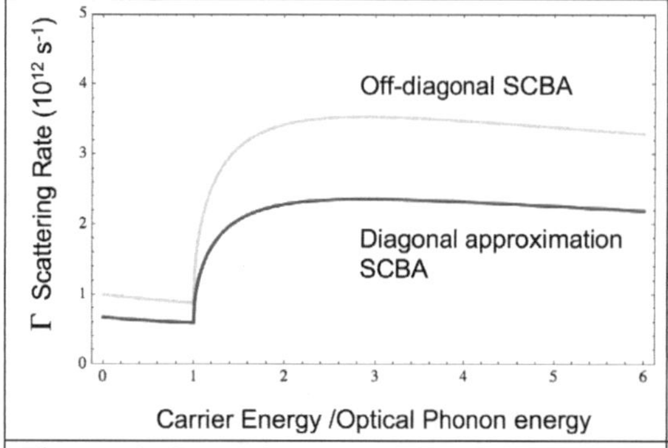

Figure 4: Scattering Rates for Polar Optical Phonon scattering

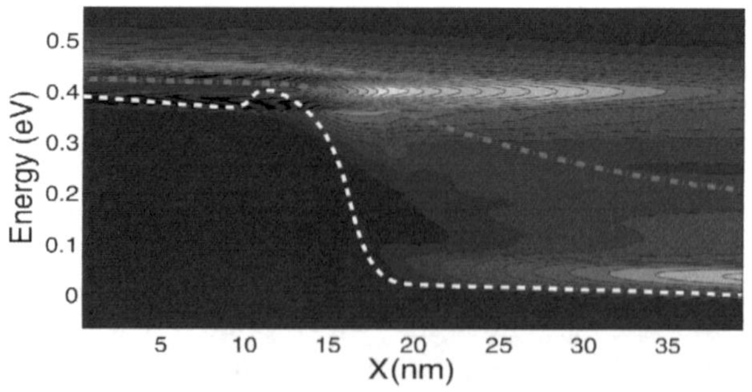

Figure 5: 3D NEGF simulation of nanowire transistor showing current spectra

The results are shown in figure 4 for n-GaAs at 300K using the data for figure 2b, for the full-off diagonal self-energy and in the diagonal approximation (in the z-variables). It is clear that diagonality under-estimates the scattering rate and indeed under-estimates the energy and momentum dissipation. Figure 5 shows our recent calculation for the current spectra in a full 3D NEGF simulation of a silicon nanowire transistor with 2.2 nm X 2.2 nm cross section and 6 nm channel length using *diagonal* self-energies. The current mean energy (red line) shows the hot electron energy relaxation in the drain. The local power dissipation is not equal to the Joule power. The white dashed line represents the first sub-band. The present paper suggests these data under-estimate the energy dissipation in the nanowire.

5. Discussion and Conclusions

The incorporation of full non-diagonal self energies in the 3D-NEGF simulation of nano-transistors presents a major computational challenge as the recursive algorithm is not immediately applicable. We are investigating re-scaling of the self-energy to a quasi-local spatial form to recover a recursive algorithm. Furthermore the approximation must be robust against instabilities arising from commensurability between the mesh spacing and the long-range quasi-periodic non-locality shown in figures 2a, 2d and fig 3b. This work demonstrates that the non-diagonality of the self-energy has a substantial effect on scattering rates and that the simple approximations for scattering kernels are strongly dependent on the value of the screening length. Confining the phonon wave vectors to the first Brillouin zone generates self-energies for which non-locality extends over a few nanometres (and thus within the spatial scale of features in extreme nano-transistors). Although local approximations to SCBA appear to be compatible with the Ward identities it does not follow that they are fully compliant with energy and momentum conservation. Locality is equivalent to isotropic scattering in the corresponding localised dimensions and under-estimates momentum and energy relaxation rates. Recent silicon nanowire devices involve strong polar coupling between channel electrons and interfacial SO phonons excited in the wrap-round dielectric: the self-energies will be highly non-local.

6. References

[1] Aldegunde M, Martinez A and Barker J R 2012 *IEEE El. Dev. Letters* **33** 194
[2] Martinez A, Seoane N, Aldegunde M, Brown A R, Barker J R and Asenov A 2011 *IEEE Trans. Electron Dev.* **58** 2209
[3] Martinez A, Bescond M, Barker J R, Svizhenko A, Anatram M P, Millar C and Asenov A 2007 *IEEE Trans. Electron Dev.* **54** 2213
[4] Aldegunde M, Martinez A and Asenov A, 2011 *J. Appl. Phys.* **110**, 094518
[5] Jin S, Park Y J and Min H S 2006 *J. Appl. Phys.* **99** 123719
[6] Banit F, Lee S-C, Knorr A, Wacker A 2005 *Appl. Phys. Letters* **86** 041108
[7] Schmielau T, Pereira Jr M F, 2009 *Appl. Phys. Letters* **95** 23111

3rd Workshop on Theory, Modelling and Computational Methods for Semiconductors IOP Publishing
Journal of Physics: Conference Series **367** (2012) 012013 doi:10.1088/1742-6596/367/1/012013

Noise Analysis of Gate Electrode Work function Engineered Recessed Channel (GEWE-RC) MOSFET

Ajita Agarwala[1,2] and Rishu Chaujar[1,3]
[1] Microelectronics Research Lab, Delhi Technological University, Main Bawana Road, Delhi-42, India.

[2] Email: ajitaagarwala90@gmail.com

[3]E-mail: rishuchaujar@rediffmail.com

Abstract. This paper discusses the noise assessment, using ATLAS device simulation software, of a gate electrode work function engineered recessed channel (GEWE-RC) MOSFET involving an RC and GEWE design integrated onto a conventional MOSFET. Furthermore, the behaviour of GEWE-RC MOSFET is compared with that of a conventional MOSFET having the same device parameters. This paper thus optimizes and predicts the feasibility of a novel design, i.e., GEWE-RC MOSFET for high-performance applications where device and noise reduction is a major concern. The noise metrics taken into consideration are: minimum noise figure and optimum source impedance. The statistical tools auto correlation and cross correlation are also analysed owing to the random nature of noise.

1.Introduction

To achieve higher speeds and packing densities, the world has witnessed the miniaturization of the basic MOS device structure. Today, CMOS technology has an established place in the design of multi-gigahertz communication circuits. This is due to the continuing down-scaling of MOS devices, which improves their RF performance characteristics. As the MOSFET is shrunk, some unwanted effects are observed such as punchthrough, hot carrier injection, noise in RF range, and dependence of threshold voltage on channel dimensions, DIBL and other short channel effect (SCEs) which affect the performance of the device in a negative manner. As MOSFET device sizes and signal levels are aggressively scaled down, the low-frequency noise (LFN) properties become increasingly important. This is because the signals are no longer significantly higher than the LFN, especially since the LFN level increases significantly as the device's size is scaled down[1,2]. As the issue of noise becomes increasingly important in deep-submicron MOSFETs, it is necessary to be able to accurately measure and model the noise parameters of MOSFET to fully characterize its noise performance.

Concave MOSFETs are known to alleviate many of the SCEs [3,4].This has been achieved by separating the source and drain (S/D) regions by a groove. Several studies have reported that the potential barrier formed at each concave corner is responsible for suppressing the SCEs, hot-carrier effects and punch-through. It is also responsible for the degradation of current driving capability and threshold voltage. The recess MOSFET, however, in conjunction with dual material gate (DMG) architecture [5,6], named as Gate electrode work function engineered recessed channel (GEWE-RC) MOSFET, as shown in figure 1, enhances drain current characteristics (shown in figure 2), average carrier velocity and suppresses SCEs, thereby proving superior to the conventional MOSFET. With DMG architecture, the step potential profile, due to different work functions of two metal gates, ensures reduction of SCEs and screening of the channel region under gate 1 from drain potential variations. Thus, the average electric field in the channel is enhanced, improving the electron velocity near the source and, hence, the carrier transport efficiency. Some past works on this device such as RF analysis[7] and linearity analysis[8] have been done before and they suggest the feasibility[9] of this model. The work has been complied in Chaujar *et al.* The GEWE-RC MOSFET [10,11] (figure 1) considered in this study also integrates the potential benefits of concave MOSFETs with DMG architecture for enhancing the noise performance of scaled devices in comparison to the conventional MOSFET.

For the purpose of the above mentioned noise analysis, structural design parameters, such as gate length, junction depth, substrate doping, gate metal work function and thickness of the oxide layer are tuned in GEWE-RC MOSFET to attain the best performance. The noise parameters examined are the minimum noise figure[12], NF_{min}, and optimum source admittance, Y_{opt}, or impedance, Z_{opt}, as they are important in the design of low noise RF circuits such as low noise amplifiers (LNA)[13] and mixers[14]. The statistical tools auto correlation and cross correlation are also analysed owing to the random nature of noise.

All simulations have been performed using ATLAS device simulation software. The models activated in simulation comprise the inversion layer Lombardi CVT mobility model along with the Shockley–Read–Hall (SRH) and Auger recombination models for minority carrier recombination. Furthermore, we adopt the hydrodynamic energy transport model which includes the continuity equations, momentum transport equations, energy balance equations of the carriers and Poisson's equation [12]. It can model the non-local transport phenomenon, and hence presents a higher accuracy than the drift–diffusion method. The quantum corrections have not been taken into account because the quantum mechanical effects become significant when the gate oxide thickness is below 30 °A or 3 nm. In our study, since the gate oxide thickness is 3 nm, the quantum corrections are ignored [15]. The specifications of the device are given in Table 1.

Table 1. Design Parameters for GEWE-RC MOSFET used in the analysis

Design Parameters	
Effective Channel Length	96nm
Groove Depth	70nm
Source/Drain Junction Depth	50nm
Negative Junction Depth	20nm
Substrate Doping	$1*10^{17}cm^{-3}$
Source/Drain Doping	$1*10^{20}cm^{-3}$
Physical Oxide Thickness	3nm
Permittivity Of Oxide	3.9
Work Function Of Gates	Of gate1=4.77eV
	Of gate2=4.1eV

Journal of Physics: Conference Series **367** (2012) 012013 doi:10.1088/1742-6596/367/1/012013

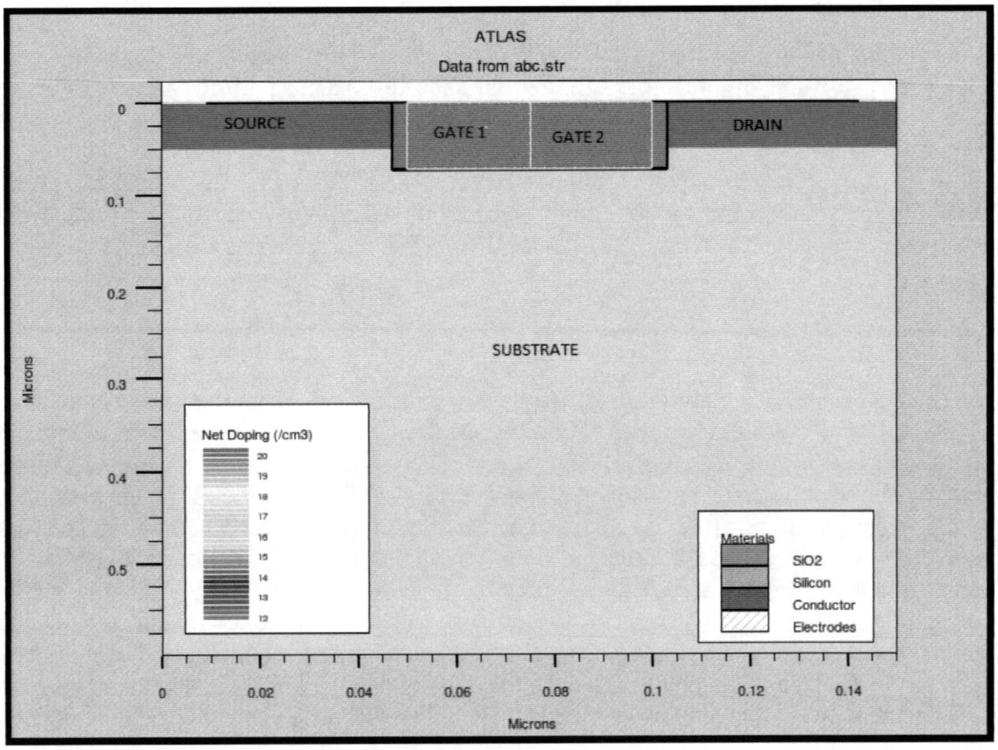

Figure 1. The architecture of GEWE-RC MOSFET.

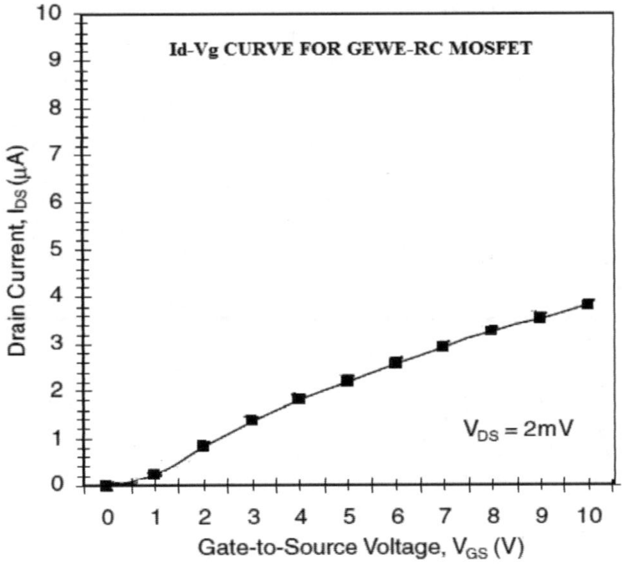

Figure 2. Id-Vg characteristics of GEWE-RC MOSFET.

2. Noise Analysis

Noise performance of a modern small-area MOS devices is dominated by random telegraph signal (RTS) fluctuations[7,12]. Their origin is the capture and subsequent emission of charge carriers at discrete trap levels near the Si–SiO_2 interface[16,17]. For deep-submicrometer devices, the number of traps with energy within a few kT close to the surface Fermi level is small[2]. Both, the number of traps and their position over the channel are random variables[16,18]. Traps located in the gate oxide near the interface to the silicon capture and reemit some of the carriers responsible for the current flowing between the source and the drain of the device[18]. Hence, the carrier transport efficiency is hampered.

2.1. Results and Discussion

2.1.1. Minimum Noise Figure. The Figure 3 explains the effect of frequency on the noise behavior of GEWE-RC MOSFET and the Conventional MOSFET designs (with the same specifications), in terms of minimum noise figure. Results clearly reveal that noise figure for the conventional MOSFET decreases with the increasing frequency but that of GEWE-RC MOSFET increases with the increasing frequency. Although, it is noteworthy that the noise figure for the conventional MOSFET and GEWE-RC MOSFET at 1000Hz is 67.6643dB and 0.000212794dB respectively. Also, it is vital to observe that the lowest noise figure shown by the conventional MOSFET is 9.18454dB only. This observation can be mainly attributed to the work function difference of the gates in the GEWE-RC architecture, due to which a step-potential is introduced in the channel[19]. There exists a screening of the channel region from the drain induced variations due to which the number of carriers entering the channel remains comparatively less varied[19]. Also, the vertical electric field is reduced due to which the trapping of the carriers near the Si-SiO_2 reduces, which results in the improved carrier transport efficiency[20].

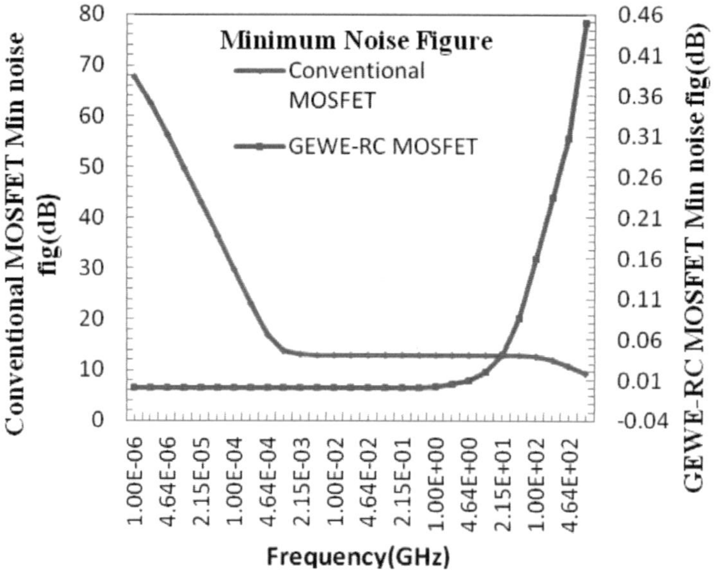

Figure 3. Minimum Noise Figure as a function of frequency.

2.1.2. Optimum Source Impedance. The Figure 4 gives the behavior of optimum source impedance ($Z_{OPT}=R_{OPT}+jX_{OPT}$) with respect to frequency. Because the oxide layer acts as a dielectric, there is essentially never any current between the gate and the channel during any part of the signal cycle[21]. As the oxide thickness is being continuously shrunk due to the scaling of the MOSFET, the MOSFET should have a large source impedance to avoid destruction by electrostatic charges. In the Figure 3 it can be vividly seen that optimum impedance for GEWE-RC MOSFET is much higher than that of the Conventional MOSFET.

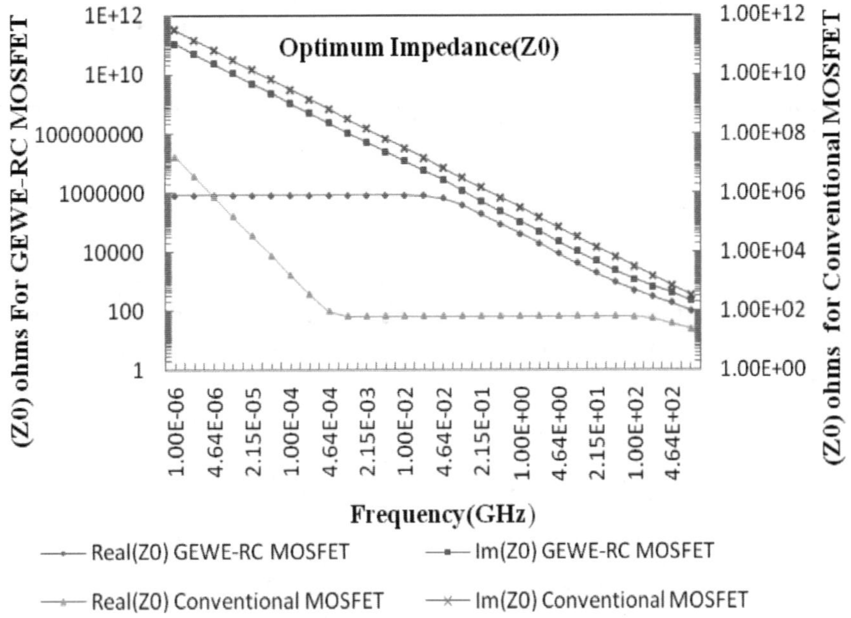

Figure 4. Optimum Source Impedance as a function of frequency.

2.1.3. Auto Correlation and Cross Correlation. The MOSFET is considered as a two port device (Figure 5) i.e. the noise induced at the gate is separated from the MOSFET and is depicted by V1 and the noise received at the output is replaced by a voltage V2[22]. Since noise is a random phenomenon, some statistical analysis is indispensable for this research. Thus, the autocorrelation and cross correlation of the voltages at the two ports of the devices are compared. From the above analysis it can be inferred that surface scattering with the Si-SiO_2 gate interface is noticeably reduced in the GEWE-RC MOSFET because of a lower vertical electric field. Therefore, isotropic scatterings present a reduced prevalence in this device[23]. Thus it can be concluded that the mean free path (λ) of carriers crossing the channel as a function of the frequency is much larger than in the conventional transistor[19]. This reduced influence of isotropic scatterings implies that the scattering mechanisms are not so effective in breaking the correlation between gate and drain current, which leads to higher cross correlation (i.e. V1.V2*) between them[23]. Figure 6 depicts such behaviour. In Figure 6, the imaginary part of cross correlation for GEWE-RC MOSFET is abruptly ending as the y axis has a log scale and log is undefined for negative values. It is notable though that the values of Imaginary part of cross correlation for GEWE-RC MOSFET abruptly finishes at 1GHz; this is due to the fact that the y axis of the graph has a log scale and the results for the Im(V1.V2*) comes out to be negative. Thus, as log is undefined for negative values, the graph for Im(V1.V2*) abruptly ends at 1GHz. Auto correlation is the cross correlation of a signal with itself. As can be seen by Figure 7, there is higher auto correlation between the input (i.e. V1.V1*) and output voltages (i.e. V2.V2*) in GEWE-RC MOSFET than the conventional MOSFET depicting lesser intrinsic noise in the former than the latter.

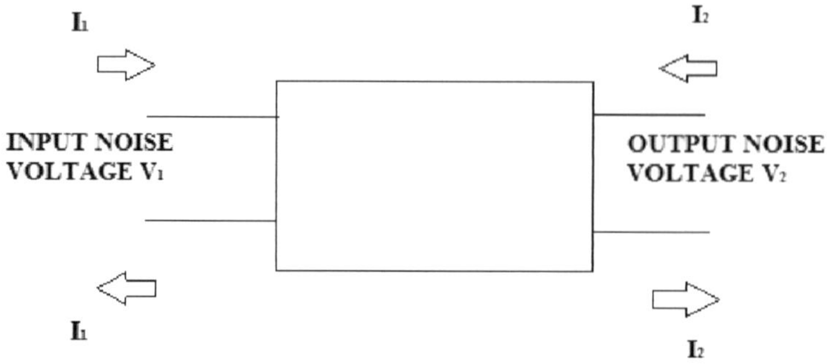

Figure 5. MOSFET as a two port device where the input noise is replace by a voltage V_1 and the the noise received at the output is replaced by V_2.

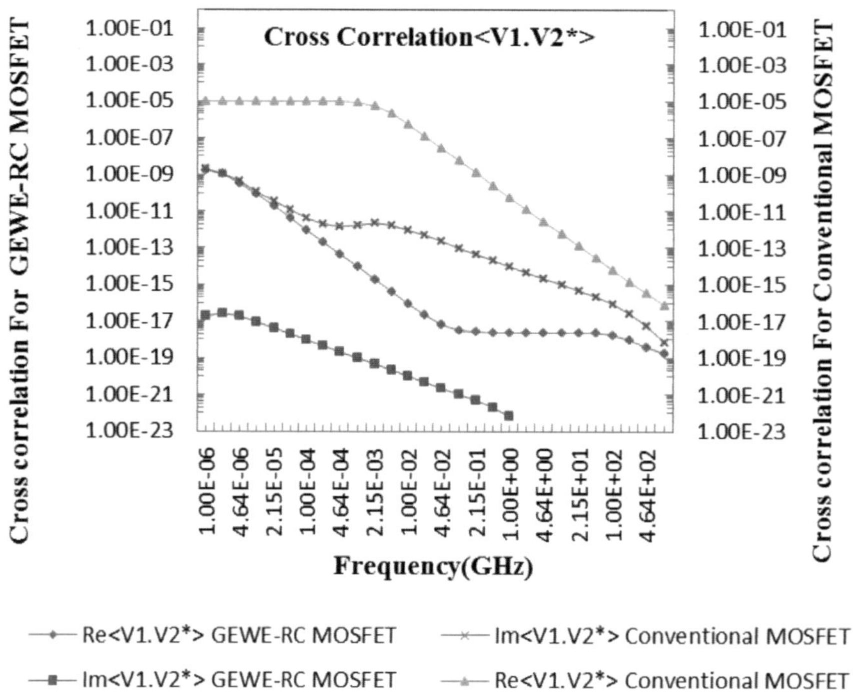

Figure 6. Cross correlation as a function of frequency.

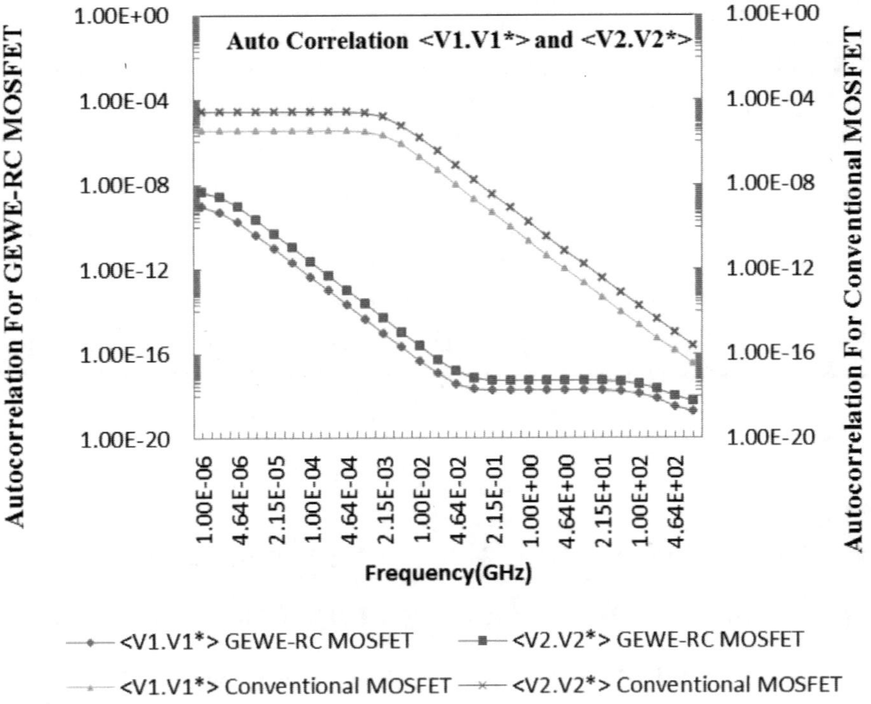

Figure 7. Auto correlation as a function of frequency.

3. Conclusion

As shown in this work, from the analysis of the microscopic noise sources and dynamic performance of the devices; GEWE-RC MOSFET exhibits superior noise performance in comparison to its conventional counterpart. It can be concluded that the reduced induced gate noise and stronger cross correlation are responsible for the noticeable improvements observed in the intrinsic minimum noise figure. Hence, proving its potency for low power, low noise and low supply voltage applications. Lower noise figure and higher optimum source impedance pertained by the GEWE-RC architecture strengthens the idea of using it for such applications, thereby giving a new opening for usage in RF applications.

4. References

[1] Brederlow R, Weber W, Schmitt-Landsidel D, and Thewes R 1999 Fluctuations of the low frequency noise of MOS transistors and their modeling in analog and RF-circuit *IEDM Tech. Dig.* 159.

[2] Sánden M, Marinov O, Deen M J, and Ostling M 2002 A new model for the low-frequency noise and the noise level variation in polysilicon emitter BJTs *IEEE Trans. Electron Devices* **49** 514.

[3] Takeda E, Kume H and Asai S 1983 New grooved-gate MOSFET with drain separated from channel implanted region (DSC) *IEEE Trans. Electron Devices* **30** 448.

[4] Sreelal S, Lau C K and Samudra G S 2002 Parasitic capacitance characteristics of deep sub micrometer grooved gate MOSFETs *Semicond. Sci. Technol.* **17** 179.

[5] Long W and Chin K K 1997 Dual material gate field effect transistor (DMGFET) *IEDM Tech.* Dig. 549.

[6] Zhou X 2000 Exploring the novel characteristics of hetero-material gate field-effect transistors

(HMGFET's) with gate-material engineering *IEEE Trans. Electron Devices* **47** 113.
[7] Chaujar R, Kaur R, Saxena M, Gupta M and Gupta R S 2008 GEWE-RC MOSFET: A solution to CMOS technology for RFIC design based on the concept of intercept point; *International Conference on Recent Advances in Microwave Theory and Applications, 2008* 661.
[8] Rishu Chaujar, Ravneet Kaur, Manoj Saxena, Mridula Gupta and R. S. Gupta 2008 Intermodulation Distortion and Linearity Performance Assessment of 50-nm gate length L-DUMGAC MOSFET for RFIC Design, *Superlattices and Microstructures* **44** 143.
[9] Chaujar R, Kaur R, Saxena M,Gupta, M and Gupta R S 2007 On-state and switching performance investigation of sub-50nm L-DUMGAC MOSFET design for high-speed logic applications *Semiconductor Device Research Symposium, International* 2007 1.
[10] Chaujar R, Kaur R, Saxena M, Gupta M and Gupta R S 2008 Laterally amalgamated DUal material Gate concave (L-DUMGAC) MOSFET *for ULSI Microelectron. Eng.* **85** 566.
[11] Chaujar R, Kaur R, Saxena M, Gupta M and Gupta R S 2007 RF-Distortion in Sub-100 nm L-DUMGAC MOSFET *14th Int. Workshop on the Physics of Semiconductor Devices* (IWPSD-2007) (Mumbai) 168.
[12] ATLAS Device Simulator, SILVACO International 2011.
[13] Shaeffer D K and Lee T H 1997 A 1.5 V, 1.5 GHz CMOS low noise amplifier *IEEE J. Solid-State Circuits* **32** 745.
[14] Karanicolas A N Dec. 1996 A 2.7-V 900-MHz CMOS LNA and mixer *IEEE J. Solid-State Circuits* **31** 1939.
[15] Lo S H, Buchanan D A, Taur Y and Wang W 1997 Quantum-mechanical modeling of electron tunnelling current from the inversion layer of ultra-thin-oxide nMOSFET's *IEEE Electron Device* Lett. **18** 209.
[16] Boutchacha T and Ghibaudo G 1998 Low frequency noise characterization of 0.18 μm Si CMOS transistors *Phys. Status Solidi, A* **167** 261.
[17] Simoen E, Dierickx B, Clayes C L, and Declerck G J 1992 Explaining the amplitude of RTS noise in submicrometer MOSFETs *IEEE Trans. Electron Devices* **39** 419.
[18] Kirton M J and Uren M J 1989 Noise in solid-state microstructures: A new perspective on individual defects, interface states and low-frequency (1/*f*) noise *Adv. Phys.* **38** 367.
[19] Chaujar R, Kaur R, Saxena M, Gupta M and Gupta R S 2008 TCAD Assessment of Gate Electrode Workfunction Engineered Recessed Channel (GEWE-RC) MOSFET and Its Multilayered Gate Architecture—Part I: Hot-Carrier-Reliability Evaluation, *IEEE Trans on Electron Devices 2008* **55** 2602.
[20] Hou F C, Bosman G, and Law M E 2003 Simulation of oxide trapping noise in submicron n-channel MOSFETs, *IEEE Trans. Electron Devices*, **50** 846.
[21] Ribes G, Mitard J, Denais M, Bruyere S, Monsieur F, Parthasarathy C, Vincent E, and Ghibaudo G 2005 Review on high-k dielectrics reliability issues *IEEE Trans. Device Mater. Rel.* **5** 5.
[22] Belostotski L and Haslett J W 2008 Two-port noise figure optimization of source-degenerated cascode CMOS LNAs, *Analog Integr. Circuits Signal Process.* **55** 125.
[23] Rengel R and Martín M J 2010 Electronic transport in laterally asymmetric channel MOSFET for RF analog applications *IEEE Trans. Electron Devices* **57** 2448.

Journal of Physics: Conference Series **367** (2012) 012014 doi:10.1088/1742-6596/367/1/012014

Optoelectronic properties of InAs/GaSb superlattices with asymmetric interfaces

Elżbieta Machowska-Podsiadło[1], Sławomir Sujecki[2], Trevor Benson[2], Agata Jasik[3], Maciej Bugajski[3] and Kamil Pierściński[3]

[1]Rzeszów University of Technology, Al. Powstańcow Warszawy 12, 35-959 Rzeszow, Poland
[2]The University of Nottingham, The George Green Institute for Electromagnetics Reasarch, University Park, Nottingham NG7 2RD, United Kingdom
[3]Institute of Electron Technology, Al. Lotników 32/46, 02-668 Warszawa, Poland

E-mail: elamp@prz.edu.pl

Abstract. In this work the optoelectronic properties of InAs/GaSb superlattices are investigated. The **k**·**p** method and four-band Kane model were used to calculate the electronic states in the structure at 0K. Superlattices with different period thickness and two types of interfaces were considered. In the calculations strain and nonparabolicity effects were taken into account. Transition energies and the corresponding cut-off wavelengths were determined.

1. Introduction

InAs/GaSb superlattices (SLs) are type-II heterostructures, which are very useful for optoelectronic applications. Their properties may be modelled with the use of the **k**·**p** method and the multi-band Kane model [1, 2, 3].

The structures are usually grown on GaSb substrates. Because of the lattice mismatch between InAs and GaSb the strain built-in the superlattice should be taken into account. In this work the strain was considered within the framework of the Bir-Pikus theory [4].

InAs/GaSb superlattices are non-common atom structures in which the symmetry is broken at the interfaces between the SL layers [5, 6]. This causes the heavy and light holes wavefunctions to mix, even in the centre of the Brillouin zone (BZ). During the growth by molecular beam epitaxy two types of interfaces may be formed in such structures, i.e. "InSb-like"and "GaAs-like"ones. Each of the interfaces has different properties and causes the coupling of holes minibands with different strength [5, 2, 6, 7, 8]. In this work SL structures with every second interface (IF) formed by In-Sb bonds were considered.

The superlattice bandgap, determined as the separation of the lowest conduction miniband and the highest miniband in the valence band, depends on the thickness of the SL layers. However, it also depends on the number and the thickness of "InSb-like"IFs in the structure [2, 6, 7].

2. Method

The electronic states in $(\mathrm{InAs})_m/(\mathrm{GaSb})_n$ superlattices were calculated with the use of the **k**·**p** method and the four-band Kane model. The method allows one to find the SL band structure in the neighbourhood of the BZ centre. The four-band Kane model describes the lowest miniband in the conduction band (CB) and the three minibands in the valence band

Journal of Physics: Conference Series **367** (2012) 012014 doi:10.1088/1742-6596/367/1/012014

(VB), i.e., heavy holes (HH) light holes (LH) and the spin-orbitally split-off (SO) band of holes. The Schrödinger equation which was solved to find the electronic states in the structure takes the form [1, 2]:

$$\begin{bmatrix} E_C + A & -\sqrt{3}\nu & \sqrt{2}U & U \\ -\sqrt{3}\nu^* & E_V - P - Q & S + iT\delta(z) & 1/\sqrt{2}S + \sqrt{2}T\delta(z) \\ \sqrt{2}U & S^* - iT\delta(z) & E_V - P + Q & \sqrt{2}Q - i\sqrt{2}Q_{kz} \\ U & 1/\sqrt{2}S^* + \sqrt{2}T\delta(z) & \sqrt{2}Q - i\sqrt{2}Q_{kz} & E_V - P - \Delta \end{bmatrix} \begin{bmatrix} \psi_{CB}(z) \\ \psi_{HH}(z) \\ \psi_{LH}(z) \\ \psi_{SO}(z) \end{bmatrix} = E \begin{bmatrix} \psi_{CB}(z) \\ \psi_{HH}(z) \\ \psi_{LH}(z) \\ \psi_{SO}(z) \end{bmatrix} \quad (1)$$

The four component ψ_{CB}, ψ_{HH}, ψ_{LH}, ψ_{SO} envelope function describes the electron and hole states in the respective minibands. The model allows one to consider the biaxial strain which is built-in the SL. In the present work this was done with the use of the Bir-Pikus deformation potentials which are included in the Hamiltonian in the terms on the diagonal. The off-diagonal blocks of the Hamiltonian account for the nonparabolicity effects which are the consequence of the narrow InAs bandgap.

The model also considers the HH and LH wavefunctions mixing at the interfaces of the structure. The strength of coupling of heavy and light hole minibands is expressed by the parameter T. The value of T depends on the type of the IF in the structure [5, 2, 8].

The effective masses of electrons (m_{el}), heavy holes (m_{HH}), light holes (m_{LH}) and the holes in the SO miniband (m_{SO}) depend on the coordinate z. In the simulations all these masses take different values in InAs and GaSb SL layers. The values of m_{HH}, m_{LH} and m_{SO} are described with the help of the Luttinger parameters. The spatially dependent effective mass of electron reflects also the nonparabolicity effect which is caused by the interaction between the valence band and the conduction band in the superlattice [1, 2].

3. Results

Calculations were performed for $(InAs)_m/(GaSb)_n$ superlattice structures with InAs and GaSb layers of equal thickness. Simulations were done for superlattices grown on (100) oriented GaSb substrates. Structures with 30 periods were considered. The electronic states were calculated at 0K and the material parameters were assumed to be as those reported in ref. [2]. A band offset energy equal to 140meV was taken into account.

Equation (1) was solved using the finite difference method. The calculations were done with the discretization mesh in which every two neighbouring nodes were separated by one monolayer (ML). Simulations were performed for the superlattices with asymmetric interfaces. For the interfaces formed by In-Sb bonds the coupling Hamiltonian H_{xy}, which determines the parameter T, was assumed to be 580meV [5, 2]. All interfaces in the structure were 1ML thick.

In the simulations the electrons and holes envelope functions were calculated in the respective minibands. Probability densities obtained for $(InAs)_{8ML}/(GaSb)_{8ML}$x30 SL structure are shown in Fig.1. The graph presents $|\psi_{CB1}|^2$ and $|\psi_{HH1}|^2$ together with the potential energy describing the bottom edge of the conduction band E_C and the top edge of the valence band E_V (see Eq.(1)). Results of calculations were obtained for the SL structure in which every second interface was formed by In-Sb bonds. It can be observed that the type of the interface is quite important for the heavy hole envelope function and it is not very significant for the electron probability density. Such effect is the result of the interaction between heavy and light holes states at the interfaces in the structure.

Interband transition energies (see Fig.2a) and the corresponding cut-off wavelengths (see Fig.2b) were calculated for $(InAs)_m/(GaSb)_m$x30 SL structures in which InAs and GaSb layers were 8, 10 and 12 MLs thick. The results show that the cut-off wavelength rises with the thickness of the SL period. This corresponds well with the experimental observation described by Rodrigues *et al.* in ref. [9]. In that work $(InAs)_m/(GaSb)_m$ SLs with m equal to 5, 8, 10 and

Journal of Physics: Conference Series **367** (2012) 012014 doi:10.1088/1742-6596/367/1/012014

15 MLs were investigated. The PL spectra recorded at 80K show that the cut-off wavelength of such structures varies from 3.9μm ($E_{\text{HH1-CB1}}$=319meV) to 8.4μm ($E_{\text{HH1-CB1}}$=147meV). Our calculations show that the absorption edge of the SL structure with m=8ML is observed at the energy $E_{\text{HH1-CB1}}$=278meV (see Fig.2a) and it corresponds to $\lambda_{\text{cut-off}}$=4.46μm (see Fig.2b). These values are in a good agreement with the absorption edge reported by E. Plis *et al.* in ref. [10]. The absorbance and PL spectra of $(\text{InAs})_{8\text{ML}}/(\text{GaSb})_{8\text{ML}}$ SL structure (measured at 72K) show that the $\lambda_{\text{cut-off}}$ of such a superlattice is equal to 4.27μm. The wavelength corresponds to the transition energy of 290.4meV. It's worth to be noted that the change of the temperature from 0 to 72K will cause the rise of the transition energy by about 6meV and simultaneously the cut-off wavelength will decrease by about 0.1μm. This signifies that the values calculated numerically and determined in the experiment are even closer. Additionally, calculations performed with the use of the pseudopotential method allowed to find almost the same value of $\lambda_{\text{cut-off}}$ as that determined in the experiment mentioned above [10].

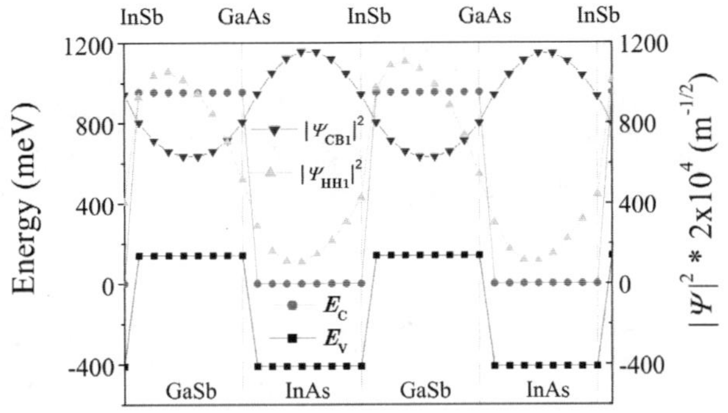

Figure 1. Square of the absolute value of the electrons and heavy holes envelope functions in $(\text{InAs})_{8\text{ML}}/(\text{GaSb})_{8\text{ML}}$x30 SL structure. $|\psi_{\text{CB1}}|^2$ and $|\psi_{\text{HH1}}|^2$ are presented together with the potential energy describing the bottom of the conduction band E_{C} and the top edge of the valence band E_{V}. Results of simulations were obtained for the asymmetric SL structure in which every second interface is formed by In-Sb bonds. Calculations were done at 0K and considered the superlattice grown on GaSb substrate.

a)

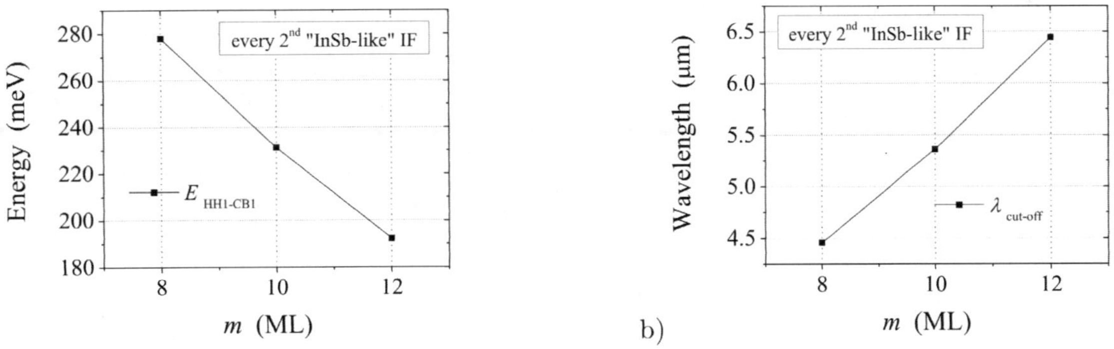

Figure 2. a) Interband transition energy and b) the corresponding cut-off wavelength calculated for $(\text{InAs})_m/(\text{GaSb})_m$x30 superlattices with asymmetric interfaces. Simulations were done for structures in which every second interface is formed by In-Sb bonds. Superlattices grown on GaSb substrates were considered. The analysis was done at 0K.

Journal of Physics: Conference Series **367** (2012) 012014 doi:10.1088/1742-6596/367/1/012014

4. Conclusion

In this work $(InAs)_m/(GaSb)_m$x30 superlattices with asymmetric interfaces were investigated. Structures with every second IF formed by In-Sb bonds were considered. Interband transition energy and the cut-off wavelength were determined for the superlattices with InAs and GaSb layers 8, 10 and 12 MLs thick.

The results of calculations are in a good agreement with the experimental observations [10, 9] and with the results of simulations undertaken using other models and methods [9, 10, 6, 8]. The bandgap of the SL structures decreases with the thickness of the SL period. The cut-off wavelength calculated for a $(InAs)_{8ML}/(GaSb)_{8ML}$ periodic structure is equal to 4.46μm. This value corresponds well with the absorption edge (4.27μm) determined for the SL structure on the basis of the absorbance and PL spectra measurements [10].

Calculations performed within the present work brings us closer to the numerical characterization of the optoelectronic properties of the strain balanced InAs/GaSb superlattices in which two types of IFs with different thicknesses are contained in the structure.

Acknowledgments

The work was supported by grant PBZ-MNiSW 02/I/2007, grant 5070/B/T02/2011/40 founded by The National Science Center and The European Cooperation in the field of Scientific and Technical Research within the COST-STSM-MP0702-8103.

5. References

[1] Liu G and Chuang S L 2002 *Phys. Rev.* B **65** 165220
[2] Szmulowicz F, Haugan H and Brown G J 2004 *Phys. Rev.* B **69** 155321
[3] Jasik A, Sankowska I, Regiński K, Machowska-Podsiadło E, Wawro A, Wzorek M, Kruszka R, Jakieła R, Kubacka-Traczyk J, Motyka M and Kaniewski J 2012 *Crystal Growth: Theory, Mechanisms and Morphology. Chapter 9 - MBE Growth of Type-II InAs/GaSb Superlattices on GaSb Buffer* (Hauppauge NY: Nova Science Publishers, Inc.)
[4] Bir L and Pikus G E 1974 *Symmetry and Strain-Induced Effects in Semiconductors* (New York: Wiley)
[5] Ivchenko E L, Kaminski A Yu and Rössler U 1996 *Phys. Rev.* B **54** 5852
[6] Magri R and Zunger A 2002 *Phys. Rev.* B **65** 165302
[7] Jasik A, Sankowska I, Pierścińska D, Regiński K and Pierściński K 2011 *J. Appl. Phys.* **110** 123103
[8] Machowska-Podsiadło E and Bugajski M 2009 *Acta Physica Polonica A Supp.* **116** 65
[9] Rodriguez J B, Christol P, Chevrier F and Joulli A 2005 *Physica E* **28** 128
[10] Plis E, Lee S J, Zhu Z, Amtout A and Krishna S 2006 *IEEE J. Sel. Topics Quantum Electron.* **12** 1269

Journal of Physics: Conference Series **367** (2012) 012015 doi:10.1088/1742-6596/367/1/012015

Empirical interatomic potential for the mechanical, vibrational and thermodynamic properties of semiconductors

U Monteverde[1], M A Migliorato[1] and D Powell[2]

[1] School of Electrical and Electronic Engineering, University of Manchester, Sackville Street Building, M13 9PL, Manchester, UK

[2] Intercollege Larnaca, 52 Famagusta Avenue, Larnaca-Dhekelia Road, Larnaca, 6019, Cyprus

Abstract. Empirical models are widely used to simulate large atomic structures where instead *ab initio* methods are not practical because of computational limitations. However models such as Tersoff potential [8], [9], Valence Force Field [13], [14] or Stillinger- Weber potential [15] have some restrictions in correctly predicting simultaneously both elastic and vibrational properties of the crystals [18]. Thus, extension of the functional form of the potentials by including further atomic interactions [20] [21] compared to the simple 2- and 3-body terms, is required. An empirical interatomic potential is proposed which represents a substantial improvement of the Tersoff potential for semiconductors modelling. The new model includes multi-bond interactions and the volume dependency by considering the tetrahedron distortions of the covalent crystal.

1. Introduction

Strain in a semiconductor changes the electronic properties hence the intrinsic stress in an heterostructure has a direct influence on device behavior [1]-[4]. In device design it is of fundamental importance to understand the relation between the strained crystalline structure and the vibrational properties [5], [6] of the material [7].

Many computational methods have been introduced to simulate crystal properties. Based on the first-principles quantum mechanics Density-Functional Theory (DFT) gives accurate descriptions but with high computational effort the size of system under simulation is limited. An easier approach is given by the tight-binding approximation, which leads to a semi-empirical method that still requires high computation. A third way to describe physical feature of a solid is by empirical potentials [8]-[15]. The great advantage of empirical methods is on the reduced calculation effort but the number of simulated properties can be limited by the functional form of the potential itself or it depends on the parameterization.

2. Conventional Tersoff potential

Based on the Morse-type potential, Abell [16] showed that the binding energy of a system can be described by considering a simple atomic pair interaction independently from the local environment. This pair potential model was further extended by multiple atomic interactions: two- and three-body terms were included in the Stillinger-Weber potential [15] as well as in the Tersoff potential. The general concept introduced by the latter model was that the tendency of two atoms to hold a bond is obstructed by their nearest atoms which exert an angular force on the bond.

The Tersoff potential, rearranged later by Smith [17], has the form:

$$V_{ij} = f_c(r_{ij})\left[V_R(r_{ij}) - b_{ij}V_A(r_{ij})\right] \tag{1}$$

$$V_R(r_{ij}) = \frac{D_e}{S-1}e^{-\beta\sqrt{2S}\left(r_{ij}-R_e\right)} \tag{2}$$

$$V_A(r_{ij}) = \frac{SD_e}{S-1}e^{-\beta\sqrt{2/S}\left(r_{ij}-R_e\right)} \tag{3}$$

where r_{ij} is the distance between the bonded atoms i and j, $V_R(r_{ij})$ and $V_A(r_{ij})$ are the repulsive and attractive potentials of the pair i-j respectively; $f_c(r_{ij})$ is a cut off function and b_{ij} is the bond-order which introduces a dependence on the local environment given by:

$$b_{ij} = \left[1 + \left(\gamma\zeta_{ij}\right)^n\right]^{-1/(2n)} \tag{4}$$

$$\zeta_{ij} = \sum_{k\neq i,j} f_c(r_{ik})g(\theta_{ijk})e^{\lambda^3(r_{ij}-r_{ik})^3} \tag{5}$$

$$g(\theta_{ijk}) = 1 + \left(\frac{C}{D}\right)^2 - \frac{C^2}{D^2 + \left(H - \cos(\theta_{ijk})\right)^2} \tag{6}$$

$f_c(r_{ik})$ is again a cut off function and $g(\theta_{ijk})$ is a function of the bond angle. The model has 10 degrees of freedom as much as the number of parameters (D_e, S, R_e, β, γ, n, λ, D, C, H) that need be fitted to properly describe the bonding strength and the behaviour of the material. In Table 1 we show the mechanical properties of Silicon simulated with the Tersoff potential along with the phonon dispersion curves in Figure 1 using a parameterization from [22].

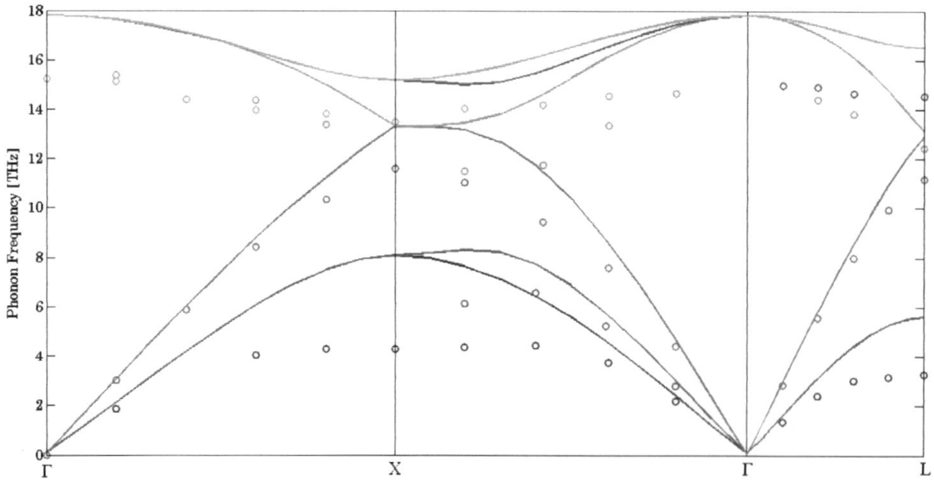

Figure 1. Phonon dispersion curves of Silicon calculated with the conventional Tersoff potential (solid lines) [22] and experimentally measured (circles) [23].

Journal of Physics: Conference Series **367** (2012) 012015 doi:10.1088/1742-6596/367/1/012015

Table 1. Mechanical properties of Silicon simulated by the conventional Tersoff potential.

Property	**Measured**	**Simulated**
E_{Coh} (eV)	-4.63	-4.63005
Lattice C. (Å)	5.431	5.43089
B (Mbar)	0.9783	0.976005
C' (Mbar)	0.509	0.511114
C_{11} (Mbar)	1.657	1.65749
C_{12} (Mbar)	0.639	0.635262
C_{44} (Mbar)	0.796	0.797153

The inconsistency of the potential in predicting at the same time the mechanical properties (Table 1) and the dispersion curves (Figure 1) is evident. The excellent estimation of the elastic constants is achieved at the expense of the phonon frequencies. The limit of the model was in part pointed out by Tersoff in his earlier publications, where the phonons overestimation was attribute to a common feature of the short-ranged models [8], [18], [19].

3. New potential

We introduced in the conventional potential model two secondary forces: a repulsive force $V_{R2}(r_{ij})$ and an attractive force $V_{A2}(r_{ij}, r_{ik})$. These forces are modulate by a 4-body term a_{jk} and a 3-body term c_{jk}, and their contribution rises when the tetrahedron is out of equilibrium. Thus, the expression of the new potential becomes:

$$V_{ij} = f_c(r_{ij})\left[V_R(r_{ij}) + a_{jk}V_{R_2}(r_{ij}) - \tilde{b}_{ij}V_A(r_{ij}) - c_{jk}V_{A_2}(r_{ij})\right] \tag{7}$$

$$V_{R_2}(r_{ij}) = B_e e^{-\rho\left(r_{ij}-P_e\right)} \tag{8}$$

$$V_{A_2}(r_{ij}, r_{ik}) = U_e e^{-\sigma\left(3r_{ij}-\sum_k r_{ik}\right)} \tag{9}$$

where the terms in tilde indicate that respect to their definitions in the Tersoff model some modifications have been introduced. The attempt here is to give to the potential a dependency upon the volume distortion and a multi-bond interaction through the terms:

$$a_{jk} = \left[1 - \frac{1}{3}\sum_{k\neq j} e^{L(r_{jk}-r_{k'k''})^2 + \eta(r_{jk}{}^2 r_{k'k''}{}^2 - r_{jk'}r_{jk''}r_{k'k'}r_{k'k''})}\right]^2 \tag{10}$$

$$c_{jk} = \left[1 - \frac{1}{3}\sum_{k\neq j} e^{W(\cos\theta_{jkk'}-\cos\theta_{eq})(\cos\theta_{jkk''}-\cos\theta_{eq})}\right] \tag{11}$$

where j, k, k" and k"' are neighbours of the central atom i and the summation is over k because the atomic distances are all defined once the edge jk of the tetrahedron is considered. The term a_{jk} is a significant term of the new model as much as the bond-order term b_{ij} of the classic potential. In b_{ij} two competing bonds (*ij*, *ik*) are considered through the angle θ_{ijk} and the difference r_{ij}-r_{ik}, whereas in a_{jk} the couple of bonds *ij-ik* compete in stretching and torsion with the opposite couple *ik"-ik"'* through the distances between bonds. The term c_{jk}, given in equation 11, plays a more effective role in the adjustment of the dispersion curves out the high symmetry points and in the Grüneisen parameters by taking into account the tetrahedron distortion through the angles of its faces $\theta_{jkk''}$ and $\theta_{jkk'''}$ having the same edge *jk*.

Compared to the original potential we have also introduced in ζ_{ij} a term depending on the angle between two faces of the tetrahedron $\theta_{jkk''k'''}$ and rewrote the angular function $g(\theta_{ijk})$ as a series of cosines which converges very rapidly to the equilibrium bond angle:

$$\tilde{\zeta}_{ijk} = \sum_{k \neq j} f_c(r_{ik}) \tilde{g}(\theta_{ijk}) e^{\lambda(r_{ij}-r_{ik})+\nu(\cos\theta_{jkk''k'''}-\cos\theta_{eq})} \quad (12)$$

$$\tilde{g}(\theta_{ijk}) = H + C\cos(\theta_{ijk}) + D\cos(2\theta_{ijk}) + E\cos(3\theta_{ijk}) + F\cos(4\theta_{ijk}) \quad (13)$$

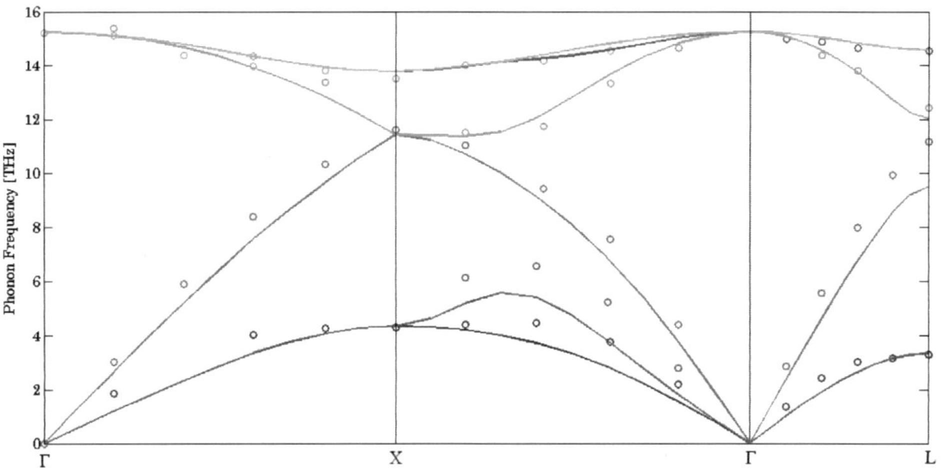

Figure 2. Phonon dispersion curves of Silicon calculated with the modified Tersoff potential (solid lines) and experimentally measured (circles) [23].

Table 2. Mechanical properties of Silicon calculated with the modified Tersoff potential.

Property	Measured	Simulated
E_{Coh} (eV)	-4.63	-4.64806
Lattice C. (Å)	5.431	5.39575
B (Mbar)	0.9783	0.946459
C' (Mbar)	0.509	0.509235
C_{11} (Mbar)	1.657	1.62544
C_{12} (Mbar)	0.639	0.60696
C_{44} (Mbar)	0.796	0.79601

Table 3. Grüneisen parameters of Silicon calculated with the modified Tersoff potential.

Grüneisen	Measured	Simulated
$\Gamma_{LO, TO}$	0.98	1.28
X_{TA}	-1.40	-1.80
$X_{LA, LO}$	0.90	1.23
X_{TO}	1.50	1.64
L_{TA}	-1.30	-1.45
L_{TO}	-1.30	-1.43

The potential model has now 11 more degrees of freedom (due to the new parameters: B_e, U_e, P_e, ρ, σ, L, η, W, v, E, F) which allow as to fit mechanical properties, phonon dispersion curves and Grüneisen parameters. The results of a simulation on Silicon with the described new potential are listed in table 2 and table 3, and the phonon dispersion curves are shown in figure 2; all simulations and parameterizations were performed on a Desktop PC using in-house software initially developed by Dr. D. Powell and Dr. M. A. Migliorato.

The phonon frequencies predicted (figure 2) at the high symmetry points, particularly for the acoustic modes, are in good agreement with the experimental data and the calculated elastic constants (table 2) are in excellent agreement compared with others empirical potentials [11] [21]. Moreover, another important result lies in the good estimation of the Grüneisen parameters (table 3), since all empirical models known fail in such estimation, mainly in the prediction of the negative values.

4. Conclusion

The model presented is able to simultaneously predict the elastic constants and phonon dispersion curves of the materials along with the Grüneisen parameters which are directly linked to the thermodynamic properties of the system under investigation. Instead, the flattening of the phonon transverse acoustic branches is poorly reproduced due to the absence in the model of the long-range atomic interactions. Although the number of parameters required by the model to fit the semiconductor's properties was increased, the computational effort was not altered since the terms introduced depend on the first nearest neighbours only.

References

[1] Choi S-M, Jhi S-H and Son Y-W 2010 *Phys. Rev. B* **81** 081407
[2] Johnson H T and Freund L B 2001 *Int. J. Solids and Structures* **38** 1045
[3] Bryant G W, Zieliński M, Malkova N, Sims J, Jaskólski W and Aizpurua J 2011 *Phys Rev B* **84** 235412
[4] Singh J 2003 Electronic and Optoelectronic Properties of Semiconductor Structures *Cambridge Univ Pr*
[5] Sui Z and Herman I P 1993 *Phys. Rev. B* **48** 17938
[6] Grundmann M, Stier O and Bimberg D 1995 *Phys. Rev. B* **52** 11969
[7] Stroscio M A and Dutta M 2001 Phonons in nanostructures *Cambridge Univ Pr*
[8] Tersoff J 1986 *Phys. Rev. Lett.* **56** 632
[9] Tersoff J 1989 *Phys. Rev. Lett.* **39** 5566
[10] Mozzone A M 1991 *Phys. Stat. Sol. (b)* **165** 395
[11] Brenner D W 1990 *Phys. Rev. B* **42** 9458
[12] Phillips J C 1968 *Phys. Rev.* **166** 832
[13] Keating P N 1966 *Phys. Rev.* **145** 637
[14] Martin R M 1970 *Phys. Rev. B* **1** 4005
[15] Stillinger F H and Weber T A 1985 *Phys. Rev. B* **31** 5262
[16] Abell G C 1985 *Phys. Rev. B* **31** 6184
[17] Smith R 1992 *Nucl. Instr. Method Phys. Res. B* **67** 335
[18] Porter L J, Justo J F and Yip S 1997 *J. Appl.Phys.* **82** 5378
[19] Powell D, Migliorato M A and Cullis A G 2006 *Physica E* **32** 270
[20] Lindsay L and Broido D A 2010 *Phys. Rev. B* **81** 205441
[21] Han P and Bester G 2011 *Phys. Rev. B* **83** 174304
[22] Powell D 2006 Thesis for the degree of Doctor of Philosophy *University of Sheffield*
[23] Dolling G 1963 *S. Ekland (IAEA Vienna)* **2** 37

AUTHOR INDEX